Laying It Down Right

T.J. Spracklin

Published by T.J. Spracklin, 2026.

LAYING IT DOWN RIGHT

First edition. January 18, 2026.

ISBN: 979-8233880315

Written by T.J. Spracklin.

Table of Contents

Introduction – Why Floors Matter

Most people never think about the floor they're walking on.

They notice the walls. They comment on the paint. They say something about the lights, the pictures, the furniture. But the thing their weight lives on—the surface their knees, backs, and shoulders feel all day—just disappears into the background.

Until it doesn't.

Until the seam pops and grabs a heel.

Until the carpet ripples and throws a shadow across a nice lobby.

Until a wrinkle catches a cart wheel, or a loose step kicks somebody in the shins.

Then suddenly, everybody sees the floor.

I've spent most of my life on my knees with a knife in my hand, fixing problems like that—and, hopefully, preventing them. Hotels, casinos, offices, showrooms, homes, garages. Some jobs you're proud to leave your card on. Some jobs you're just proud you survived.

This book is about the difference.

It's not a glossy manufacturer's brochure, or a training manual written by somebody who's only ever touched a trowel in a photo. It's a working installer's view of how to lay carpet right—the kind of right that lasts past the final draw, past the warranty, past the first round of furniture and the second round of management.

If you're a helper just starting out and trying not to screw up the baseboard, a working installer who wants to tighten up your game, or an owner, GC, or facility manager who actually cares what's under people's feet, this book is for you.

We're going to cover tools, layout, seams, subfloor prep, stretching, stairs, problem jobs, and the stuff nobody puts in the glossy instructions—the politics, the pressure, the shortcuts you should never take, and the ones that might save you.

You won't find many pictures here. You will find real decisions from real jobs, the "why" behind doing it a certain way, and the little habits that separate a hack from a craftsman.

You don't have to be famous to be a master installer. You don't need a logo on your truck or a hundred thousand followers. You just need a standard you don't back away from when the room gets tight, the GC gets loud, or your knees start talking.

If that sounds like the kind of installer you want to be—or the kind of work you want on your project—keep reading. Let's lay it down right.

Part I – Foundations

Chapter 1 – The Invisible Trade

There's a funny thing about carpet.

If you do it wrong, everyone sees it.

If you do it right, nobody thinks about it at all.

Ask most people to describe a beautiful room and they'll talk about the walls, the windows, the light fixtures, the furniture. They'll tell you about the view, the art, the curtains. If they mention the floor at all, it's usually just, "Yeah, it looked nice."

They don't see the seam that's running right under their feet.

They don't think about the tension holding that stretch.

They don't notice the prep work hiding under that pad.

Until something fails.

Until a wrinkle throws a shadow across the lobby.

Until a loose step bites a heel.

Until a seam grabs the wheel of a luggage cart and snaps the front desk out of its trance.

Then suddenly, everyone sees the floor.

I've spent most of my life on my knees with a knife in my hand, fixing those moments—or preventing them. Hotels, casinos, offices, schools, homes, and garages. Some jobs you're proud to leave your card on. Some jobs you just hope nobody pins on your name.

This book is about the difference.

It's not written from a boardroom, or a marketing office, or a manufacturer's sample room. It's written from the slab up—from the view of someone who has actually scraped the glue, pulled the stretcher, and gone back on the callback.

If you're just getting into the trade, I want you to understand something right away:

You're signing up for a craft that is **most successful when it's invisible.**

That messes with some people. They want applause. They want the customer to stand in the middle of the room and say, "Wow, who installed this?" And sometimes that happens.

Most of the time, it doesn't.

Most of the time, what you get is silence.

People walk across your work without tripping.

They push carts and chairs over your seams without thinking.

They run vacuums and buffing machines, wear out the furniture, repaint the walls, and remodel the reception desk—all while your floor quietly does its job.

If you did it right, they will never know your name.

Do it wrong, and they'll learn it fast.

That's the invisible trade.

What carpet really does

BEFORE WE TALK ABOUT knives and stretchers and seam tape, you need to understand what carpet really does for a space. It's more than "soft" or "pretty."

A good carpet installation:

- **Protects people**
  It reduces slips, cushions falls, and smooths transitions between surfaces. That matters in hotels, hospitals, schools, and homes more than most people realize.
- **Controls sound**
  Walk into a fully tiled lobby and then walk into a carpeted corridor. The difference in echo, footsteps, and background noise is your work at play.
- **Shapes how a room feels**
  Underfoot comfort, warmth, perceived quality—all of that changes when you go from hard, cold surfaces to a well-installed carpet.
- **Guides movement**
  Patterns, borders, and seam locations help steer people from entrance to desk, from hallway to room, from stair to landing. You're not just laying material; you're quiet traffic control.

When you think of carpet this way, "good enough" stops being good enough. A wrinkle isn't just ugly—it changes how the whole room works.

Learning to see like an installer

REGULAR PEOPLE WALK into a room and see color and furniture.

An installer walks in and starts scanning automatically:

- Where's the main traffic path?
- Where does the light hit strongest?
- Where are the trouble spots in the substrate?
- Where can I hide a seam—and where will it shout at me later?

If you stay in this trade long enough, there's usually one moment where your brain flips. You stop seeing "floor" and start seeing seams and work.

It might happen:

- In a hotel hallway where the pattern breaks three inches off center.
- On a stair run where every tread is cut just a little different.
- In a dentist's office, a bubble grabs your foot every time you cross the room.

Once you see it, you can't unsee it.

You'll find yourself looking down in restaurants. You'll spot seams in airports while everyone else is staring at their phone. You'll check stair nosings with your boot as you walk down. That's not you being picky—that's you getting trained.

If you're new, lean into that. Use every floor as a classroom:

- Ask how they handled the seams.
- Look at where the wrinkles showed up first.
- Notice how pattern falls at doorways, columns, and transitions.

Every bad floor is a free lesson someone else paid for. Every great floor is a standard you can steal.

Craft vs. commodity

OUT IN THE WORLD, CARPET installation often gets treated like a commodity.

On a bid list, you're a number.

On a spreadsheet, you're a line item called "carpet labor."

On a big job, you might be one of five prices in a column that someone scans in five seconds.

Glue is glue, right?

A stretch is a stretch.

"If you guys are all licensed and insured, why shouldn't I just pick the cheapest one?"

If you let that thinking sink into your own head, you're done. You'll start believing you *are* just a number, and that your only move is to race to the bottom and pray your body holds out.

You can't control how every GC, buyer, or homeowner thinks. What you *can* control is:

- How you see your own work
- How you talk about it
- What line you absolutely will not cross, even when money is talking loud

If you see yourself as "just a carpet layer," you'll chase square yards and fight every penny until your knees or your back give out.

If you see yourself as a **craftsman**, you still must deal with pricing, scheduling, and paperwork—but you're building a body of work you're not ashamed to stand on later.

A commodity job says: "Just get it covered."

A craft job says: "Let's make sure this still looks and feels right when the next guy peels back the corner in fifteen years."

The real customer you're working for

ON PAPER, YOUR CUSTOMER might be:

- A GC
- A facility manager
- A building owner
- A homeowner who just wants their living room redone

But there's another customer you're working for, and they're tougher than all of those:

The next installer who sees your work.

Someday, somebody is going to:

- Pull up that carpet
- Scrape that pad
- Look at your seams and stretch
- See your prep and your patchwork

They will know instantly what kind of installer you were that day.

They may never know your name.

They'll still know if you cared.

That may sound dramatic, but it's real. We've all had that moment on demo:

- "Damn, this guy knew what he was doing."
- Or: "Whoever did this should've been banned from touching a stretcher."

One of the goals of this book is simple:

When the next installer sees your work, I want them to nod, not swear.

The deal you're really making

EVERY TIME YOU TAKE a job, you're not just trading time for money.

You're making a deal:

- You give your labor, your knowledge, your body, your patience.
- In return, you get a check—and your name attached to that floor, whether anyone sees it or not.

If you're in this for more than six months, this trade will shape you:

- The way you solve problems under pressure
- The way you handle mistakes
- The way you talk to other trades, to clients, to your own crew
- The way your back and knees feel when you get out of bed

That's all part of it. You don't get to opt out.

So from the very start, decide what kind of deal you're willing to make:

- Are you okay leaving behind "good enough as long as nobody looks too close"?
- Or do you want to be the person people call when they actually care how it turns out?

This isn't about perfection. Nobody hits 100%. The job site is messy, schedules are tight, product shows up wrong, and you will make mistakes.

But there's a big difference between:

- Doing the best work you can in a tough situation, and
- Cutting corners so hard you're basically gambling with someone else's building.

One is part of the trade. The other will haunt you.

Where we're going from here

THIS FIRST CHAPTER wasn't about technique on purpose.

Before we talk about knives, stretches, seams, and stairs, you need to be clear about what you're actually doing when you walk into a room with tools in your hand.

You're stepping into an invisible trade:

- If you succeed, most people will never say anything.
- If you fail, they'll feel it every day.

The chapters ahead will get into:

- Tools that feel like extensions of your body
- How to read a room and a subfloor before you unload the truck
- Planning cuts and seams so you're not just reacting
- Subfloor prep that actually holds
- Seams, patterns, stretching, and stairs
- Working in occupied spaces without making enemies
- Dealing with GCs, chain accounts, and callbacks
- Protecting your body and building a name that actually means something

But all of that sits on top of this:

If you do your job right, most people will never know your name.

Do it anyway—and do it in a way you're proud of.

Job Note – The Floor I Couldn't Ignore

I STILL REMEMBER THE first time I couldn't "turn off" seeing floors.

We were checking into a hotel as a family. I wasn't working there, wasn't bidding the job, wasn't supposed to be thinking about carpet at all. I was just carrying bags, hunting for the ice machine, and trying to be on vacation.

Then we turned down the main corridor and I saw it instantly.

The pattern was drifting a little more at every doorway. Seams sat dead center at doorways in the traffic path. Wrinkles were just starting to shadow under the wall sconces, like the carpet was already tired of holding on. My wife didn't say a word. She was focused on room keys and giftshop hours. But my installer brain was running in the background, doing the math on where the crew must have started their layout, what they'd prioritized, and how fast that stretch must have gone.

I didn't complain to the front desk. I didn't ask who installed it. I just logged it—another quiet reminder of what I don't want my work to look like five months after I leave a job.

You'll have your own moment like that, if you haven't already. When it happens, don't just grumble about how bad somebody else's work is. Ask yourself, "What would I have done differently?" Then do that on your next job.

Checklist – Are You Treating This Like a Craft?

USE THIS AS A QUICK gut-check at the start of your career, and again every few years.

On your last few jobs:

- Did you walk the entire space before unloading, or just start dragging tools in?
- Did you think about traffic and light when placing seams, or just "where the roll fits"?
- Did you speak up when the substrate was bad, or quietly cover it and hope for the best?

- Did you check your own work at the end of the day, or just pack up when the clock said so?
- Do you remember the last time you changed a method because you learned a better way?

If most of your answers are "no," you're treating this like a paycheck.

If your answers are mostly "yes," you're on your way to being a craftsman.

Either way, the rest of this book is here to help you move in the right direction—one room, one corridor, one invisible floor at a time.

Chapter 2 – Tools of a Master Installer

Every trade has its tools, but carpet installation is different in one important way:

Most of your work lives in that thin line between **sharp** and **dull**, **tight** and **loose**, **clean** and **cluttered**.

You can be the smartest installer in the world, but if your knife is dull, your stretcher's missing pieces, and your tools are buried under trash in the van, you're going to do hack work without meaning to.

This chapter is about:

- What you need
- What you don't
- And how to treat your tools like part of your body, not random junk in a bucket

Buy Once, Cry Once

THERE ARE TWO KINDS of tool stories in this trade:

1. **"I bought the good one and it's still with me."**
2. **"I tried to save 20 bucks and it cost me a whole day."**

Early on, you're broke and every tool feels expensive. Cheap gear is tempting. And to be fair, not everything has to be top-of-the-line. But there are certain tools where "buy the real thing" will save you:

- Time
- Pain
- Embarrassment on the job
- And sometimes, losing a whole client

Any tool that touches:

- **Your body** (kneepads, shoes, pouch)
- **Your primary cut** (knives, blades)
- **Your tension** (stretcher, kicker)

...is worth buying right.

You don't have to get everything perfect on day one. Think of it like this:

Start with a solid basic kit.

Then, every time a tool hurts you, slows you down, or fails under pressure, ask:

"Is this the last time I want to feel this, or am I okay repeating this lesson?"

That question should guide your upgrades.

Essential Hand Tools

THESE ARE THE TOOLS that live in your hand all day. You don't need twenty gadgets; you need a small group of reliable, well-maintained core tools.

Knives

YOUR KNIFE IS YOUR paintbrush, your scalpel, and your signature.

You'll want:

- A **hook-blade utility knife**
    - For cutting backing without diving into pad or subfloor
    - Great for rough cuts, trimming off rolls, cutting pad
- A **straight-blade carpet knife**
    - For precision cuts, seams, and fine trimming
    - Lives close to your heart, metaphorically and sometimes literally in your chest pocket

Key habits:

- Change blades **more often than you think you need to**
    - Dull blades make you push harder
    - Pushing harder makes you slip
    - Slipping gets you stitches or a chewed-up edge
- Keep spare blades **organized and dry**
- Never toss blades under the pad or in a loose bucket
    - Use a blade box or old container
    - Don't leave razor traps for your future self

A sharp knife is safer than a dull one if you respect it.

Trowels or Rollers

TROWELS AND ROLLERS control how your adhesive hits the floor.

You'll need:

- A **basic notched trowel** in the sizes your adhesives call for
- A **low nap roller** in the size your adhesive calls for
- Choose witch of **different types and sizes** if you're doing:
    - Different backings
    - Different adhesives
    - Different subfloor porosity

Rules of thumb:

- Don't "eyeball" it when the manufacturer calls for a specific notch
- Replace trowels when the notches wear down and start starving the floor
- Clean them before the glue turns into concrete

A worn-out trowel is a sneaky enemy. It'll have you under-spreading adhesive without realizing it.

Stair Tools, Hammers, and Friends

YOU'LL USE THESE MORE than you think:

- **Stair tool / bolster**
  - For tucking carpet into grippers and gully
  - For dressing edges at metal and baseboard
  - For installing stairs
- **Hammer**
  - Light enough to swing all day
  - Heavy enough to persuade tackstrip and transitions
- **Pry bar / flat bar**
  - For demo, pulling strip, lifting metal, sneaking under trim
- **Chisels / 4" razor scrapers**
  - Cleaning edges, stubborn glue, or junk in corners

You don't need a whole mechanic's chest. You do need:

- Tools that feel good in your hand
- Steel that won't bend or shatter under normal use

If you dread reaching for a tool because you know it's junk, it's time to replace it.

Layout and Measuring Tools

MEASURE WRONG, AND no tool in the world can save you.

Tape Measures

YOU WOULDN'T THINK a tape measure could ruin your day. It can.

Good habits:

- Buy a tape that **you can read easily**

    - If you're squinting or guessing, you're going to round wrong
- Develop a standard way of reading: .0, 3", 6", 9"
- converts on the calculator .25, .50, .75
    - Always speak the full dimension out loud (e.g., "Eleven feet, three")
    - Repeat it back when someone reads it to you

Double-check critical dimensions:

- Hallway widths
- Room lengths
- Stair tread depths
- Landings and closets

One bad number can cost you a whole piece of carpet.

Chalk Lines and Lasers

STRAIGHT LINES ARE cheaper than crooked ones.

- **Chalk line**
  - For snapping centerlines in rooms and corridors
  - Use chalk color that cleans or hides well
- **Laser (optional but powerful)**
  - For checking square rooms vs. reality
  - Help line up patterns and seams in long runs

You don't have to get fancy. But you **do** have to:

- Establish reference lines
- Stop trusting the walls to be square

Squares, Straightedges, and Story Sticks

- **Drywall square / framing square**
  - Checking corners, marking cuts
- **Straightedge 3' and 6'**
  - Guiding cuts and door seams
  - Checking for bows and humps

THESE TOOLS SAVE YOUR mental energy. Use them.

Seaming and Stretching Tools

THIS IS WHERE THE DIFFERENCE between "installer" and "hack" really shows.

Seaming Irons and Tape

THE SEAM IRON MIGHT not run every day, but when it does, you want it to be right.

- **Seam iron**
    - Heats evenly, v-groove for extra center glue, adjustable thermostat
    - Cord in good condition
- **Seam tape**
    - Right tape type and quality for:
        - Commercial or Residential
        - Cushion-backed carpets
    - Don't cheap out—seam tape is holding your floor together

Fundamentals:

- Keep the iron **moving at a steady pace, one iron length at a time.**
- Apply both carpet edges together – No gaps – No overlaps
- Use tractor to press hotmelt glue into carpet
- Use **seam weights** or boards to flatten and hold the seam as it cools

Let seams set properly before stretching or traffic

When you rush a seam, the floor remembers.

- **Seam tractor**
    - Solid carpet tractor on Plush (Rub with hand)
    - Spike carpet tractor on Glue-down

Knee Kicker vs. Power Stretcher

THIS IS THE BIG ONE.

Knee kicker:

- Good for:
    - Positioning carpet into strip for set walls
    - Working in closets, small cuts, detail spots
- NOT a substitute for:
    - Proper stretching in a full room or corridor

Power stretcher:

- Non-negotiable for **real** stretch-in work
- Allows:
    - Proper tension across the room
    - Controlled stretch direction
    - Safer, repeatable results

Components to keep in good shape:

- Head (teeth sharp and clean)
- Poles (not bent, joints not slipping)
- Tail block / end bracing

If you're still trying to stretch whole rooms with a kicker because "that's how the old guys did it," understand this:

Most of the wrinkles you're re-stretching in old buildings were done by someone who thought the same way.

Be the installer who uses the power stretcher—not the one calling themselves back in five years.

Prep and Demo Tools

YOU CAN'T LAY A GOOD floor on top of bad junk. The tools you use to **take things up** matter just as much as the ones you use to put things down.

Scrapers

YOU'LL USE:

- Hand scrapers
- Long-handle scrapers
- Possibly powered scrapers on bigger jobs

Use them to:

- Remove old adhesive ridges
- Knock down debris
- Clean edges and corners

Sharpen or replace blades when they start "chattering" or skipping. Dull scrapers are like dull knives—dangerous and useless.

Grinders and Surface Prep Machines

NOT EVERY INSTALLER owns grinders, but you should at least understand them.

- Walk-behind or handheld grinders:
    - Flatten high spots
    - Remove stubborn glue or coatings
- Always pair with:
    - Good dust collection
    - Proper PPE (mask, eye protection, hearing protection)

If you don't own these tools, you should at least know:

- Who to call
- How to ask for the right prep
- How to charge for it if you're managing the work

Vacuums and Cleaning

A GOOD SHOP VAC IS non-negotiable.

- Clean subfloor = better adhesive bond
- Less dust = better working conditions for you and everyone else
- Clean edges = better-looking stretch and seams

It's not "extra fancy" to vacuum before you glue. It's professional.

Safety and Body-Saving Gear

YOU ONLY GET ONE BACK, one neck, and two knees. The trade doesn't care; your tools don't care; the schedule doesn't care. You have to care.

Kneepads

CHEAP KNEEPADS ARE a fast way to hate your life.

Look for:

- Thick, supportive padding that doesn't bottom out
- Straps that don't cut off circulation or chafe behind your knee
- Non-marking durable outer shell or surface

Replace kneepads when:

- They're permanently flattened
- They feel like bare floor through the pad
- The straps are shot and keep slipping

If you've been thinking, "They're still fine," but your knees say otherwise, listen to your knees. Future you will be glad you did.

Footwear

YOU'RE ON YOUR FEET and knees all day.

Good work shoes:

- Support your arches
- Cushion your heel
- Grip well on dusty concrete and smooth backs

Avoid:

- Worn-out sneakers with no tread
- Heavy boots that make you clumsy and wears you out mid-day

Gloves, Eyes, Ears

YOU WON'T ALWAYS WEAR them, but you'll wish you had them on the day something goes wrong.

- **Gloves**
    - For demo unknown old carpet funk, scrap handling, metal
- **Safety glasses**
    - For grinding, scraping, installing strip, and surprise shrapnel
- **Ear protection**
    - For grinders, saws, and nosey onlookers

You're already in a tough trade. No need to make it harder by going half-protection.

Tool Care and Organization

A PRO INSTALLER AND a hack can own the **same** tools and get completely different results, just based on how they care for them.

Blades and Edges

- Change knife blades before they fight you
- Replace worn trowels and scraper blades
- Keep a small container for all your old blades

IF YOU'RE FORCING CUTS, you're either:

- Using the wrong blade
- Or using the right blade way past its time

Keeping Your Kit Tight

EVER WORKED OUT OF someone's van and thought:

"I could lose a whole small child in this mess"?

Don't be that guy.

Basic organization:

- Tool bag or box with **assigned spots** for essentials
- Small parts (blades, screws, anchors) in labeled containers
- Daily habit: quick reset at the end of each job

A clean setup:

- Saves time looking for tools
- Reduces lost tools
- Sends a message to clients and GCs that you're not a clown show

Truck and Trailer Habits

YOU DON'T HAVE TO BE Instagram-level tidy. But:

- Keep trash out of your tool area
- Secure heavy tools so they don't become missiles
- Have a routine for loading and unloading so you don't forget critical gear

Nothing makes you feel dumber than:

- Driving to a job
- Walking into a big empty space
- Realizing the stretcher is still in the garage

Job Note – The Cheap Tool That Cost a Day

I ONCE WATCHED A GUY show up to a commercial job with a brand-new bargain-bin stretcher. On paper it had everything: head, poles, tail block. It looked fine. Shiny, even.

On the very first stretch across the room, the poles flexed like wet spaghetti, the joints slipped under pressure, and the head jumped a tooth and chewed the backing. We spent the next hour tightening, re-tightening, and babysitting a tool that was obviously not up for the job.

By midday, we had burned time fighting the stretcher, burned extra energy just trying to get tension, and ended up borrowing an old, beat-up stretcher from another crew on site just to finish the room. Did we save money on the purchase? Sure. On the job? Not even close.

The lesson was simple: if a tool must hold your work together, don't cheap out. You don't have to buy the most expensive name in the catalog, but you do need gear that was built for real use, not staged photos. Every time you're tempted by the cheapest version of a critical tool, remember that stretcher and ask yourself if you really want your name riding on something that folds under pressure

Checklist – Building Your Starter Kit (and Pro Kit)

USE THIS AS A QUICK reference. You can add brands and preferences as you go, but the categories don't change much.

Starter Kit (Helper / New Installer)

- Hook blade utility knife + spare blades
- Straight blade knife + spare blades
- Basic notched trowel (per adhesive spec)
- Stair tool / bolster
- Wall trimmer
- Light hammer
- Pry bar / flat bar
- Tape measure (readable, not junk)
- Chalk line
- Simple straightedge
- Hand scraper
- Shop vac (or access to one)
- Decent kneepads
- Work shoes with grip

Pro Kit (Lead Installer / Owner)

EVERYTHING IN THE STARTER kit **plus**:

- Multiple trowel notch sizes
- Seam iron
- Quality seam tape (right types for your work)
- Seam roller + seam weights / boards
- Knee kicker (in good shape)
- Power stretcher with full pole set and tail block
- Long-handle scrapers
- Access to grinders / prep machines (owned or rented)
- Organized tool bag or box
- Blade disposal container
- Extra PPE (glasses, gloves, ear protection)

Maintenance Habits

- Change blades before they force you to push
- Clean trowels before glue turns to rock
- Vacuum your work area before spreading adhesive
- Reset your tool setup at the end of each job
- Replace kneepads and worn tools before they injure you

YOU DON'T BECOME A master installer just by owning nice tools. You become one by:

- Choosing the right tools,
- Using them the right way,
- And taking care of them like they're part of your body.

In the next chapter, we'll take these tools into the room and talk about something most people rush through:

Reading the space and the subfloor before you ever cut a piece.

That's where your tools stop being random gear and start becoming part of a plan

Chapter 3 – Reading the Room (and the Subfloor)

Most installers lose more time and money by **not looking** than by not knowing.

They don't walk the whole job before unloading.

They don't really look at the subfloor before they start scraping.

They don't think about light, traffic, or the way the room works.

Then later, they're shocked when:

- A seam shadows under a window
- A bubble shows up over a crack
- A pattern walks across the corridor
- Or the GC asks, "Why didn't you say something before you started?"

This chapter is about the most underrated skill in the trade:

Seeing the job before you start the job.

If you learn to read a room and its subfloor, you'll make better choices about layout, seams, prep, and even pricing. You'll also have fewer "How did this happen?" moments, because you'll have seen most of it coming.

Step 1 – Walk the Space Before You Touch a Tool

ON A NEW JOB, THE WORST thing you can do is:

- Back the truck up
- Drag all your tools in
- Start demo or cutting

...without a full walk-through.

Make it a rule:

Before I unload, I **walk**.

Walk every room, every hallway, every stair, every transition area you're responsible for. While you walk, you're not just looking—you're asking questions in your head.

Look up, out, and down:

- **Up:**
    - Where are the lights?
    - Where are the windows and how does sunlight hit the floor?
    - Any low-hanging fixtures, sprinklers, or weird ceilings?
- **Out:**
    - Where will people walk the most?
    - Where do they enter and exit?
    - Where are desks, counters, furniture, or displays likely to sit?
- **Down:**
    - What's the current floor doing?
    - Any obvious cracks, dips, humps, or old patch lines?
    - What's at the thresholds—metal, tile, concrete, transitions?

This first walk is where you start to spot:

- Danger zones
- Opportunities
- And lies the drawings might be telling

Don't rush it. Ten extra minutes now can save hours later.

Step 2 – Understand How the Room Will Be Used

CARPET DOESN'T LIVE in a vacuum. It lives on the floor.

When you read a room, ask: *"What kind of life is this floor going to have?"*

Traffic Patterns

WHERE WILL MOST FEET go?

- In a hotel:
  - Lobby: doors → front desk → elevators
  - Corridors: elevator → rooms
- In an office:
  - Main path: entrance → reception → workstations
  - Secondary: break room, conference rooms
- In a house:
  - Living room paths, stairs, hallway to bedrooms

The main traffic path is where:

- Wear shows first
- Wrinkles will be felt most
- Seams will take the worst abuse

You want to:

- Avoid seams right in the heaviest path if possible
- Make sure tension and prep are bulletproof where traffic is strongest
- Plan your stretch pattern to "pull against" expected movement, not give in to it

Point Loads and Abuse Zones

SOME SPOTS GET MORE than footsteps:

- Front desks and check-in counters
- Copy machines and printers
- Reception chairs with metal feet
- Heavy furniture that never moves

These areas may need:

- Extra prep attention
- Different seam placement
- Or sometimes, a different material altogether (and you should say so)

Spill and Dirt Zones

WHERE WILL JUNK HIT the floor?

- Entrances without proper mats
- Vending or coffee areas
- Break rooms and lounges
- Near restrooms

Knowing this helps you:

- Push for proper walk-off mat coverage if you can
- Plan darker or more forgiving carpet tones / patterns in those areas
- Set realistic expectations with the client

You're not just installing carpet; you're installing **under a whole lifestyle.**

Step 3 – Light: Your Friend and Your Enemy

LIGHT IS ONE OF THE biggest reasons seams and wrinkles "show" or "hide."

On your walk, really look at:

- **Natural light:**

    - Large windows
    - Glass doors
    - Skylights
- **Artificial light:**
    - Overhead fixtures
    - Wall sconces
    - Spotlights

Ask yourself:

- "If there was a seam here, would the light catch it?"
- "If the carpet buckled slightly here, would it shadow like a mountain range?"

Rules of thumb:

- Seams running **perpendicular** to strong light (like across a bank of windows) will show more than seams running **parallel** to the light.
- Wrinkles near straight light sources (like a row of can lights) cast shadows that make small problems look big.
- Pattern misalignment is more visible in bright, even light than in dim or broken lighting.

You can't always put seams exactly where you want. But if you ignore how light hits the floor, you're setting yourself up to be the last guy everyone blames when the sun comes up.

Step 4 – Reading the Subfloor

ONCE YOU'VE READ THE "life" of the room, it's time to read what's under it.

If the old floor is still down, you're reading clues.

If it's already stripped, you're staring at the truth.

Concrete vs. Wood

CONCRETE:

- Look for:
    - Cracks (hairline vs. major movement)
    - Control joints (saw cuts)
    - Patches and old leveler
- Ask:
    - Are these cracks stable, or is something still moving?
    - Are we spanning joints that might open or shift?

Wood subfloors:

- Look for:
    - Soft spots
    - Squeaks
    - Loose or uneven panels
    - Signs of moisture or rot
- Ask:
    - Is this deck solid enough to hold tension?
    - Do we need to screw down squeaky or loose areas?

You don't have to be an engineer. But you **do** have to:

- Notice what's stable and what isn't
- Flag anything that doesn't look or feel right
- Understand that bad movement under good carpet still equals bad floor

High Spots and Low Spots

WALK THE FLOOR WITH:

- Your eyes

- Your feet
- A straightedge when needed

High spots:

- Show up as:
  - Humps
  - Ridges under old flooring
- Cause:
  - Rocking furniture
  - Telegraphed lines
  - Hard-to-stretch areas

Low spots:

- Show up as:
  - Birdbaths
  - Shallow dips
- Cause:
  - Hollow feel
  - Wrinkles forming around low zones

You're asking:

- "What needs to be ground down?"
- "What needs to be filled?"
- "What can I reasonably correct with patch, and what needs more?"

Ignoring this step is like trying to paint over a wall full of holes and expecting it to look flat.

Step 5 – Old Flooring as a Crime Scene

IF YOU'RE TEARING OUT existing carpet, pad, vinyl, or tile, treat it like a case file.

What you see coming up tells you how the last person worked—and what you're about to inherit.

Look at:

- How the seams held up
- Where wrinkles formed
- How the subfloor looks
- Whether tac-strip is spaced right, straight, and secure
- Whether stairs failed at the nosing, the riser, or the landing

Don't just rip and toss. Take 30 seconds at key spots:

- "Why did this fail here?"
- "What did they do that I won't?"
- "What held up surprisingly well?"

The previous installer left you a free training course. Might as well attend.

Step 6 – Spec vs. Reality

ON PAPER, EVERY PROJECT looks:

- Square - Level - Dry - Perfectly coordinated

On site, things are:

- Out of square - Off by inches or feet - Half demo'd - Full of surprises

If you have drawings or a spec, use them—but don't marry them.

Compare:

- **Plan:** Where seams are drawn, what materials are listed, indicated transitions.
- **Reality:** Where walls are, what's built, what's been changed in the field.

Ask:

- "Does this seam location make sense with the actual room and traffic?"
- "Did someone move walls or doors after the drawing was made?"
- "Is the specified transition realistic with the materials actually here?"

If something on the plan is clearly wrong for the **real** room:

- Take notes
- Take pictures
- Bring it up **before** you cut

This is where you protect yourself from:

"That's not what we wanted." or "Why did you do it like this?"

Step 7 – Knowing When to Speak Up

READING THE ROOM AND subfloor isn't just for your own head. It should shape the conversations you have **before** the job starts.

You don't need to turn into a drama queen every time you see a hairline crack. But you **do** need to speak up when:

- The subfloor is clearly out of tolerance
- There's obvious moisture damage
- The planned seams are a guaranteed problem
- The substrate is unsafe (soft spots, rot, crumbling concrete)
- There are design expectations that carpet alone can't solve

How you speak up matters.

Bad way:

"This is all wrong. You guys screwed this up."

Better way:

"Hey, before we start, I want to show you a couple of things I'm seeing so we're on the same page about what this floor can and can't do."

Use photos. Use simple language. Offer options:

- "We can install over this as-is, but here's what I expect will happen."
- "If we patch/grind this, it will add time and cost, but it'll perform better long term."

You don't have to be a hero. You just must give the client a **clear choice**.

Step 8 – Building Your Pre-Job Habit

THE WHOLE POINT OF reading the room and subfloor is building a **repeatable habit**.

On every job—big or small—your first 15–30 minutes should follow a rough pattern:

1. Walk the entire space.
2. Think about traffic, light, and use.
3. Read the subfloor: high/low spots, cracks, soft areas.
4. Study existing flooring (if any) for failure clues.
5. Compare what you see to the plan or spec.
6. Decide what you need to fix
7. Find your starting point
8. Only then do you unload tools and start demo or prep.

This doesn't have to take all day. On a one-room job, it might be five minutes. On a hotel floor, it might be half an hour.

But skipping it is like driving with your eyes half-closed. You might still get there, but you're increasing the odds of hitting something you didn't need to hit.

Job Note – The Bubble I Built Myself

YEARS AGO, I RUSHED a prep job in a small office. The old carpet had been pulled up, adhesive ridges were everywhere, there were a couple of low spots near the entrance, and the schedule was tight with the client breathing down our necks.

We scraped. We vacuumed. But instead of grinding down the worst high spots and patching the little birdbath by the door, I told myself, *"Pad and stretch will hide it. We don't have time for more."*

On install day, it looked fine.

Six months later, I got the call:

"There's a bubble by the door. The carpet feels weird when you step on it."

We went back and peeled it open. The low spot at the entrance had let the pad and carpet settle strange, and the old adhesive ridges we'd only half-scraped had turned into little speed bumps under the pad. Traffic and time had exposed exactly what I didn't want to see.

The client didn't care that we'd been rushed. They just wanted it to be fixed.

I ended up cutting the carpet back, doing the patch and grind I should've done the first time, and reinstalling on my own time. The lesson was simple: the ten minutes I "saved" on prep cost me hours on a callback. The thing that bothered my feet on day one became the thing bothering their feet on day 180.

Since then, when a floor doesn't feel right under my boots before the carpet goes down, I listen.

Checklist – Pre-Job Walk-Through

KEEP THIS IN YOUR HEAD—OR print it and toss it in your truck. Before you start a new job, ask yourself:

1. Space & Use

- Have I walked every room, hall, and stair I'm touching?
- Do I know where the heaviest traffic will be?
- Do I see "abuse zones" (front desks, vending, carts, entrances)?

2. Light

- Where does natural light hit the floor hardest?
- Where do artificial lights make shadows that could show seams/wrinkles?
- Am I planning seams with light direction in mind?

3. Subfloor

- Concrete vs. wood vs. "other"—do I know what I'm on?
- Have I noted obvious high and low spots?
- Any cracks, soft spots, or damaged areas that could move under the carpet?

4. Existing Flooring (if present)

- What failed on the old floor—and why?
- Are there old patch lines or adhesive ridges I need to address?
- Did the last installer's choices teach me anything I can use?

5. Plan vs. Reality

- Have I compared the drawing/spec to what's actually built?
- Are planned seam locations realistic with real walls and traffic?
- Did anything change (moved walls, added doors) that affects layout?

6. Communication

- Is there anything I need to bring up before I start?
- Do I have photos or notes for any problem areas?
- Have I given the GC/owner clear choices where conditions are questionable?

If you can honestly check most of those boxes, you're not just "showing up to lay carpet." You're stepping into the job as someone who sees the whole picture.

That's what separates a mechanic from a master.

Part II – Core Skills

Chapter 4 – Measuring and Planning Like a Pro

Most bad installs start long before the glue hits the floor.

They start on a scrap of paper with numbers that weren't checked...

or a room that was never really measured...

or a hallway where someone said, "We'll figure it out when we get there."

By the time you're fighting a seam in the wrong spot, staring at a piece that came up six inches short, or trying to stretch into a closet that should have been cut separate... it's too late. The carpet's cut. The roll's light. The truck is far away.

This chapter is about moving the work **upstream**, so the hardest part of the job happens with:

- A tape measure
- A pencil
- A quiet brain

Instead of:

- A knife
- A stretcher
- A panicked phone call

Good installers cut on purpose.

Hack installers cut and then try to explain.

We're aiming for the first one.

Why Planning on Paper Is Cheaper Than Fixing on Site

BEFORE WE GET INTO the how, let's talk about *why* this matters so much.

Every cut you make is a decision you can't take back.

- Cut too short, and you're ordering more material or stealing from somewhere you shouldn't.
- Cut with seams in the wrong places, and you're making your life harder for the rest of the job.
- Cut without respecting pattern, and you're stuck wrestling it in the field.

Spending 30–60 minutes to really measure and plan:

- Saves full days of rework
- Prevents emergency "fix it" tricks you're not proud of
- Makes your crew faster because everyone knows the plan

Think of your cut sheet as the **map**.

Once you're on site, you want to follow the map—not make up directions with the clock ticking.

Step 1 – Measuring the Space: No Guesswork

YOUR TAPE WILL LIE to you if you use it lazy.

Measure the Longest and Widest Points

ROOMS ARE ALMOST NEVER perfectly square.

If you measure:

One side length and one side width and call it done, you're asking to get bit.

Instead:

- Measure **both** lengths (left wall and right wall)
- Measure **both** widths (top wall and bottom wall)

If they don't match:

- Use the **largest** measurement for your planning
- Note where the extra distance is (room flares on one side, etc.)

In corridors:

- Measure the **narrowest** point and the **widest** point along the run
- Note any pinch points (columns, alcoves, jogs)

In stair runs:

- Measure **every** tread and riser if they look even slightly inconsistent
- At least measure top, middle, and bottom—if they differ, treat them all as suspects

Diagonals and Squareness

IF A ROOM REALLY MATTERS—LIKE a banquet hall or a patterned lobby—check squareness:

- Measure diagonally from corner to corner both ways
- If diagonals are way off, the room is out of square

You can still do the job. But you need to:

- Know where to "cheat" pattern
- Decide where you want things to look right (main view) and where you're okay hiding adjustments

Write It Down Immediately

BAD HABIT:

- Measure a room
- Think "I'll remember this"
- Get distracted
- Mix up numbers

Better habit:

- Measure
- **Say the number out loud**
- **Write it down immediately** on your pad or cut sheet

Treat the paper as the truth, not your memory.

Step 2 – Cut Sheets and Diagrams: Your Map on Paper

A CUT SHEET IS WHERE you turn raw measurements into actual pieces of carpet.

It doesn't have to be pretty. It must be **accurate**.

Basic Cut Sheet Layout

AT MINIMUM, YOUR CUT sheet should show:

- Room or area name (Lobby, 301–310 Corridor, Suite 215)
- Finished length and width
- Pile direction / arrow (or pattern direction)
- Cut size you plan to take off the roll (with waste and stretch allowance)

Example:

Room 204

- Finished: 21'6" x 14'5"
- Cut1: 12'0" x 14'9" (add Minium 3" to every cut)
- Fill: 10' x 14'9"

The point is to decide *beforehand*:

- How many pieces you're cutting
- Where seams will land
- Which direction pile or pattern will run

Diagrams Matter More Than Words

ALONG WITH NUMBERS, include a simple sketch:

- Rough room outline

- Mark doors, windows, and major features
- Draw seam lines
- Draw arrows showing pile/pattern direction

Nobody expects a blueprint. But a quick sketch:

- Makes it easier to explain the plan to your helper or crew
- Gives you something to double-check in the room before cutting
- Shows you problems on paper instead of in carpet

Step 3 – Pattern Repeats and Roll Allocation

IF YOU'RE WORKING WITH patterned carpet, planning becomes even more important.

Patterns do not care about your feelings, your schedule, or how much material the client ordered. They only care about **repeats**.

Learn the Repeat

FIND OUT:

- Pattern repeat **length** (inches or feet along the length)
- Pattern repeat **width** (across the roll)

This should be:

- On the label
- On the spec
- Or measure it yourself from the carpet

Repeat sizes affect:

- How much extra you need at each seam
- How you align pattern in hallways, rooms, and at doorways
- How the pattern will "walk" across a series of rooms

Add Pattern Matching Allowance

FOR EACH PIECE THAT will join another:

- Add extra length for pattern match at the seam (could be a full repeat or more)
- Consider how many repeats you want visible in a room or corridor

Don't just cut everything to finished length plus a random number.

Conscious pattern planning might mean:

- Extra material
- Smarter seam locations
- Less headache fighting pattern in the field

Roll Allocation

ONCE YOU KNOW YOUR pieces, you need to think about how they come out of the roll.

- Total yardage vs. rolls they ordered (one roll? multiple?)
- Whether multiple rolls are from the same dye lot
- Whether you should keep particular areas on the **same roll** (e.g., main lobby and main corridor)

If you don't think about this ahead of time, you may end up:

- Having to seam pieces from different rolls where it really shows
- Short on material for a key area because you wasted it on a minor room

Use your cut sheet to:

- Group pieces smartly

- Mark which roll they should come from if needed

Step 4 – Seam Diagrams: Where the Battle Is Won

SEAM PLACEMENT IS WHERE your measurement and planning skills really pay off.

Principles of Seam Placement

WHEN DRAWING SEAMS, think about:

- **Traffic:**
    - Avoid seams dead center in high-traffic paths if possible
    - Better near walls, under furniture lines, or in low-traffic margins
- **Light:**
    - Avoid seams perpendicular to strong light sources when you can
    - Be extra careful with seams toward big windows or glass walls
- **Substrate:**
    - Don't place seams over known cracks, joints, or problem spots
    - If you must, at least be conscious of what you're spanning
- **Pattern:**
    - Align seams where pattern mismatch is least likely to scream at you
    - Use joints, columns, or slight jogs to disguise pattern shifts if needed

Corridors and Long Runs

IN HALLWAYS:

- Decide if you're:
    - Running carpet lengthwise (fewer seams, may need multiple

    pieces end-to-end)
    - Or crosswise (more seams, different visual appearance)

End seams:

- Stagger them where possible
- Try to land them near doorways or natural breaks

Side seams:

- Keep them out of the heaviest traffic when you can
- Avoid "railroad tracks" where two seams run long and parallel in full view

Rooms with Attachments

- Alcoves
- Bay windows
- Connecting closets

THINK ABOUT:

- Whether to cut those areas as separate pieces
- Or keep them as part of the main piece to avoid extra seams

Step 5 – Waste vs. Efficiency: Knowing When to Let It Go

EVERY INSTALLER WANTS to "use every scrap." Every owner wants zero waste. Reality lives in the middle.

Your job is to:

- Avoid stupid waste
- Accept smart waste

Stupid waste is:

- Cutting wrong because you didn't measure
- Throwing away usable pieces because your plan changed five times
- Not nesting cuts on the roll when you easily could have

Smart waste is:

- Extra material to allow proper pattern match
- A bigger cut piece that lets you avoid a terrible seam location
- A small area that gets its own piece instead of forcing a risky stretch

When planning:

- Look for ways to nest smaller pieces around larger ones on your cut sheet
- Consider rotating pieces when the pattern or pile direction allows it (and when spec permits)
- Don't cut a piece so tight that there's zero room to trim—give yourself some breathing space

It's cheaper to throw away a few inches of carpet

than to throw away your reputation over a bad seam.

Step 6 – Communication Before Cutting

SOMETIMES YOUR PLAN will bump up against:

- The spec
- The designer's expectations
- The GC's cost fears
- The material order they've already placed

This is when you talk **before** you cut.

Examples:

- If the spec shows a seam dead center under the worst light:
    - "I can install it as drawn, but the seam will absolutely show here. If you're okay with that, I'll proceed. If not, I recommend we move it to here."
- If the material ordered is exactly enough with no room for proper pattern match:
    - "With this yardage, we can install, but we can't guarantee full pattern match everywhere. If you want better alignment, we need X more yards."
- If you've identified subfloor issues that affect cut sizes:
    - "This room has a big out-of-square condition. I can cheat the pattern toward the main entrance view, but I'll need to plan cuts around that. Just making sure we're on the same page."

You're not asking for permission to do a good job. You're:

- Showing that you've thought it through
- Giving them clear trade-offs
- Protecting yourself in case they pick the cheaper or quicker option

Step 7 – Quick Planning Examples

Example 1 – Basic Bedroom

ROOM: 11'6" X 14'8", 12' wide goods.

Plan:

- Cut: 12'0" x 15'0" (to allow 3" to 6")
- Always measure in 3", 6", 9", 0"
- Drop room no fill seam

Simple, clean, no extra seams. Waste is cheap here.

Example 2 – Hotel Room with Entry and Closet

ROOM: MAIN AREA + NARROW entry + closet off to side.

Options:

1. **One large piece** with pattern consideration, cuts folded into entry/closet
    a. Pros: fewer seams
    b. Cons: heavier to handle, more planning for pattern
2. **Main room piece + separate closet piece**
    a. **Pros: easier handling**
    b. **Cons: seam at closet, possible waste**

Measure carefully, diagram both options, choose what balances effort, seams, and material.

Example 3 – Long Corridor (Patterned)

CORRIDOR: 5' WIDE, 80' long, patterned 12' goods.

Plan:

- Decide direction: running lengthways or crossways
- Map out repeats and where seams (if any) will fall
- Consider starting at the visual anchor (elevator bank or lobby end)
- Plan cuts so pattern “lands” correctly at main views and door openings

No one wants to find out halfway down the hall that you’re drifting off pattern at the main entrance.

Job Note – The Hallway That Taught Me to Look Down

I ONCE WALKED ONTO a hotel renovation where the plan called for one long corridor of patterned carpet—clean, simple, "no seams in the middle," according to the paperwork. On the drawing, everything looked straight. The pattern ran neatly down the page, seams hidden in all the right places. On paper, it was perfect.

The real hallway was another story. The corridor jogged around an elevator core. The walls weren't parallel. Three structural columns were tucked into odd little alcoves. And every afternoon, the sun blasted in from one side through a long glass wall, throwing hard light across everything.

If I had trusted the drawing and just started running pattern from one end, we would've landed a pattern break right at the jog, parked seams under the worst possible light, and lined up columns and doorways in a way that made every repeat look crooked. It would have been one of those jobs you can't walk without seeing everything that's wrong.

Instead, we walked the space first and let the real hallway, not the plan, tell us what to do. We snapped a control line that made sense from the way people actually saw the corridor. We shifted our starting point about eight inches so the pattern "cheated" in a way only we would ever notice. We moved a seam into a less obvious spot near a natural change in lighting.

It cost us maybe twenty or thirty extra minutes with a chalk line and a few conversations. It saved us from living with a nasty pattern break and ugly seams for the entire life of the floor.

The GC never knew how close that hallway came to being a disaster. That's the invisible part of the trade. But we knew.

Checklist – Planning Before You Cut

USE THIS BEFORE YOU slice a single piece off the roll.

Room / Area Measurements

- Measured both lengths (long sides)
- Measured both widths (short sides)
- Used the **largest** dimension where needed
- Checked diagonals or obvious out-of-square conditions in key areas

Cut Sheet & Diagrams

- Each area has a clear label (Room 204, 3rd Floor East Hall, etc.)
- Finished sizes written down
- Planned cut sizes include trim/stretch allowance
- Pile/pattern direction marked with arrows
- Simple sketch with doors, windows, and seams drawn

Pattern & Roll Planning (if patterned)

- Pattern repeat length and width known
- Extra material allowed for matching at seams
- Plan accounts for how pattern will fall in main views
- Rolls assigned for key areas if multiple dye lots involved

Seams & Layout

- Seam locations decided based on traffic, light, and substrate
- Avoided seams in the highest wear paths where possible
- Avoided seams over obvious cracks or problem areas where possible
- Corridor seams staggered logically, not in a straight "railroad" if you can help it

Waste vs. Efficiency

- Looked for smart nesting of cuts on the roll
- Accepted some waste to avoid terrible seam locations
- Avoided razor-thin cuts that leave no room to trim

Communication

- Any layout or spec conflicts identified
- Photos/notes taken of major issues (out-of-square, bad subfloor, etc.)
- Talked to GC/owner about major trade-offs **before** cutting

If you can hit most of those points consistently, you're no longer just "good with a knife."

You're someone who can see the job in your head and on paper **before** it happens on the floor.

That's what separates an installer who survives from one who builds a reputation.

Next up, we're going to take all that planning and aim it at the thing that quietly makes or breaks almost every job:

Subfloor prep.

That's where jobs are won or lost long before anyone sees the carpet.

Chapter 5 – Subfloor Prep: Where Jobs Are Won or Lost

Everybody wants to talk about the pretty stuff:

- Patterns
- Seams
- Stretch
- Stairs

But most jobs are won or lost on something way less glamorous:

What you decide to cover up.

The subfloor is the floor your floor stands on. If that layer is dirty, loose, cracked, humped, or soaked... it doesn't matter how good your cut and stretch are. The problems will find a way through.

The worst part?

By the time you see the problem, you're usually done:

- The carpet's in.
- The room is furnished.
- The schedule is "closed."
- And somebody says, "I think there's something under the carpet."

This chapter is about learning to respect what's underneath:

- How to read it
- How to fix what you can
- How to say "No" when you should
- And how to avoid being blamed for things you didn't cause—but would have to live with

The Truth About Prep

HERE'S THE UNROMANTIC reality:

- Subfloor prep is hard to sell.
- It's hard to bill.
- And it's the first thing people want to cut when money or time gets tight.

But:

- Wrinkles, buckles, hollow spots, and telegraphing?
- A lot of that traces back to prep.

As an installer, you're stuck between:

- The material manufacturer (who says "perfect substrate or no warranty")
- The GC/owner (who says "just get it done")
- And your name, sitting right in the middle

The goal is not perfection on every job. The goal is:

Do the best prep you can inside the real-world constraints—

without crossing the line where you know it's wrong.

To do that, you need a process.

Step 1 – Strip It Down and Clean It Right

YOU CAN'T SEE THE TRUTH of a subfloor until you get the junk off it.

Remove Old Pad and Tackstrip

ON TEAR-OUT:

- Get **all** pad up—no loose chunks, no stray staples
- Remove old tackstrip where:
    - It's rotten or water-damaged
    - The pins are rusted away
    - It's in the wrong place for the new install
    - It's crushed or loose

When you pull strip:

- Watch for:
    - Nails left behind
    - Splintered concrete or wood
    - Chunks torn out around fasteners

Replace strip where needed. Don't try to "save" dead strip—it'll only betray you later.

Pull Fasteners and Sharp Surprises

WALK THE FLOOR LIKE you're barefoot, even if you're not:

- Exposed nail heads or staples
- Broken screws
- Random metal or debris

Pound, Pull or grind them down.

Anything that can:

- Telegraphed through
- Catch a knife
- Or cut the backing

...needs to go.

Deal with Old Adhesives

DIFFERENT GLUE, DIFFERENT headaches:

- Old carpet adhesive ridges
- Vinyl or VCT glue
- "Mystery mastic" from who-knows-when

Basics:

- Scrape off loose, brittle ridges
- Knock down high spots
- Vacuum up dust and chunks

If you suspect:

- Asbestos in old tile or adhesive
- Hazardous materials
  → **Stop and follow local rules.**
  This is where you let testing and abatement pros do their job. No floor is worth your lungs or your license.

Vacuum and Degrease Where Needed

DUST IS A SILENT ENEMY:

- It weakens adhesive bond
- It gets into seams

- It makes leveling and patching fail

Before you call the floor "ready" to patch or glue:

- Run a shop vac over everything
- Pay extra attention to edges and corners

If there are obvious contaminants (oil, heavy dirt, grease):

- Clean or degrease
- Don't just go over it and hope

Clean isn't glamorous, but it's part of craftsmanship.

Step 2 – Flat vs. Level: Know the Difference

- Manufacturers usually require **flat**, not perfectly level.
- But if it's neither flat nor level, *you're* the one who gets blamed.

What Is "Flat" in Our World?

FLAT MEANS:

- No sudden humps or dips
- A smooth, gradual plane
- No ridges or birdbaths that make the carpet bridge or sag

- No more than a certain variation (like 1/8" or 3/16") over a certain distance (like 6' or 10')

Lay a straightedge or long level in different directions:

- If you see daylight under the middle → low spot
- If the straightedge rocks → high spot

Level Is Not Always Required

IF A ROOM IS OUT-OF-level but mostly flat:

- Your job is usually to deal with the flatness first
- Level complaints are a different conversation

Step 3 – High Spots and Low Spots: Fix What Counts

YOU CAN'T FIX EVERYTHING on every job. But there are some things you **must** address.

High Spots (Humps and Ridges)

HIGH SPOTS CAUSE:

- Rocking furniture
- Telegraphed lines under carpet
- Hot spots where tension fights itself

Options:

- **Grind** them down (concrete)
- **Plane** or sand them (wood)
- **Chip** loose humps and re-patch as needed

Don't:

- Just stretch carpet hard over a spike or hump and hope it disappears
- Try to "float" huge high spots with feather patch alone

Low Spots (Birdbaths and Sags)

LOW SPOTS CAUSE:

- Hollow feel underfoot
- Wrinkles forming around the depression
- Adhesive pooling if you're gluing goods

Options:

- Patch with proper floor patching compound
- For larger areas, consider self-leveling underlayment (if budget and scope allow)

Blend patches out so you're not creating new ridges with your "fix."

Step 4 – Patching and Smoothing: Making It Workable

YOU DON'T NEED TO BE a full-time floor-leveler. But you do need basic patching skills.

Choose the Right Material

KNOW THE BASICS OF the products you use:

- Patch compounds designed for:
    - Feathering and smoothing
    - Filling small holes or divots
- Self-levelers designed for:
    - Larger, deeper areas
    - Pouring and letting them find their own level (within reason)

Read the bag, or talk to someone who has. Two important questions:

- Is it approved under carpet and pad / glue-down?
- What are the **dry times** before you can install?

Mix It Right

SLOPPY MIXING = WEAK patch.

Habits:

- Measure water and powder reasonably in bucket
- Mix to the recommended consistency (not soup, not dry putty)
- Let it rest for a minute if the product calls for it
- Remix and apply smoothly with a flat surface trowel

Too wet:

- Weak, chalky, cracks easily

Too dry:

- Doesn't bond well, hard to trowel flat

Feather Those Edges

A PATCH THAT LOOKS like a speed bump is not a good patch.

When you trowel:

- Press the edges thin
- Extend feather edges farther than you think you need
- Aim for no big "lip" where patch meets existing subfloor

Feel it with your hand once it's dry:

- If you can feel a ridge easily, the carpet will feel it more

Respect Dry Time

IF THE BAG SAYS:

- "Ready to cover in X hours"

...and you're walking and installing in half of that, you're gambling.

You might get away with it sometimes. But rushed patch can:

- Break down
- Powder
- Release from the subfloor under traffic

Do what you can to:

- Patch early in the day or the day before when possible
- Build dry time into your plan instead of hoping no one notices

Step 5 – Moisture: Know Enough to Not Be Stupid

THIS BOOK ISN'T A FULL moisture manual, but you need a street-level understanding.

Carpet is more forgiving than some coverings, but it's not magic. Moisture can still:

- Break down adhesive
- Promote mold in pad
- Rust strip and fasteners
- Cause smells and health complaints

Obvious Red Flags

ON YOUR PREP WALK, pay attention to:

- Dark, damp-looking areas on concrete
- Efflorescence (white powdery mineral deposits)
- Recent floods or water stains on walls and base
- Musty smell that doesn't make sense

If you see those:

- Ask about leaks or known water issues
- Note it in writing
- Consider whether moisture testing has been done (often the GC or owner's responsibility)

Simple Field Checks (Awareness, Not Official Tests)

YOU MIGHT SEE SOMEONE do a simple plastic test:

- Tape down clear plastic over suspect areas
- Check later for condensation or darkening

This is **not** a replacement for proper testing, but it can confirm that something is off.

When in doubt:

- Don't promise your floor will fix a moisture problem
- Don't warranty what the substrate can't support

Your name should not be the one absorbing decades of building mistakes.

Step 6 – Edge and Transition Prep

PEOPLE DON'T JUST TRIP in the middle of the room. They trip at the edges.

Doorways and Thresholds

AT DOORS:

- Make sure subfloor height changes are handled
- Grind or patch thresholds so there's no sharp step for carpet to bridge
- Plan where tackstrip or transition metal will sit

If a door swings over the carpet:

- Consider clearance now, not later
- Undercutting door jambs and casing looks cleaner and lets carpet tuck better

Tackstrip Placement

- Too close to the wall → no proper gully, hard to tuck
- Too far from the wall → carpet falls short, loose edge, ugly tucks
- Crooked or wavy → your stretch fights all the way along

WHEN RESETTING OR INSTALLING new strip:

- Keep a consistent gap (usually about the thickness of the carpet)
- Make sure strip is solidly fastened (no rocking, no loose pins)
- Avoid placing strip over soft or crumbling edges

Transitions to Other Floors

AT TRANSITIONS (TILE, LVT, wood, etc.):

- Check height differences
- Plan which side gets feather patch or shim to reduce trip edge
- Place metals straight and secure in solid substrate

A perfectly stretched carpet that ends in a sloppy transition still feels like hack work.

Step 7 – Knowing When to Say "No"

THIS IS THE HARDEST part of subfloor prep:

Knowing when it's not your job to "make it work" anymore.

Some conditions are **not** yours to fix:

- Structural movement or major cracks
- Active leaks or ongoing moisture issues
- Rotten wood subfloors
- Dangerous or suspect old materials (asbestos, serious mold, etc.)

Your responsibility is to:

1. **See it**
2. **Document it** (photos, notes)
3. **Tell the right person** (GC, owner, property manager)
4. **Explain the risk** of installing over it

If they say:

- "Do it anyway, we'll take responsibility"

...then you:

- Get it in writing if you can
- Decide if you're okay putting your name on that job anyway

Sometimes the strongest prep move you can make is walking away.

Job Note – The Room That Moved After I Left

I DID A STRETCH-IN job once in a condo that looked perfect when we packed up.

It was one of those days where everything felt tight on time. Furniture had to be shifted room by room, the elevator window was short, and the building manager was hovering with that "Are you done yet?" energy. We scraped, padded, set tackstrip, kicked, stretched, trimmed, tucked. When we took the tools out, the carpet looked clean and flat. I took a couple of quick photos, shook hands, and left.

Two weeks later, the call came in.

"There's a ripple in the living room," the owner said. "It wasn't there at first. Now it's right where we walk."

We went back. The wrinkle wasn't huge, but it was there, running across the traffic path, catching light from the sliding doors just enough to show itself. The furniture had been moved around. The room had seen some heat and humidity changes. And the truth was simple: we'd under-stretched it because the day was tight and the pressure was on.

Nothing had "failed." Nothing was defective. The carpet did exactly what carpet does when there's just enough slack for life to pull it loose—we gave it room to move, and it took it.

We ended up shifting furniture again, popping a few door casings, re-stretching the room properly, and putting everything back together on our own time. The owner was polite about it, but I knew we were there because I'd accepted "looks good for now" instead of "this will still be flat when the seasons change."

That job burned a simple rule into my head: if you're stretching, *stretch*. Use the right tools. Take the time. Don't let somebody else's schedule talk you into leaving tension on the table. Floors don't care how tired you were or how many jobs you had that week—they only show whether you did it right.

Checklist – Subfloor Prep: Did You Really...?

USE THIS CHECKLIST before you call a subfloor "ready."

Removal & Cleaning

- All old pad removed (no chunks left stuck to the floor)
- Bad or loose tackstrip removed and replaced where needed
- Nails, staples, screws, and sharp debris pulled or ground down
- Old adhesive ridges knocked down in main areas and seams
- Floor vacuumed, especially edges and corners

Flatness

- Checked key areas with a straightedge (entries, main paths, seams)
- High spots identified and addressed (ground, planed, or chipped)
- Low spots identified and patched where they matter

Patching & Smoothing

- Used appropriate patching products for the job
- Mixed patch to proper consistency (not soup, not dry crumb)
- Feathered patch edges smoothly into existing subfloor
- Allowed patch to dry per product guidelines (as much as schedule allowed)

Moisture & Damage

- Looked for signs of moisture (dark spots, efflorescence, musty smell)
- Noted any water-damaged areas or suspect substrates
- Raised serious moisture/rot/asbestos concerns with GC/owner instead of ignoring them

Edges & Transitions

- Tackstrip placement checked for consistent gully and solid grab

- Door thresholds checked and prepped (no big lips or dips)
- Transitions to other flooring planned and subfloor prepped accordingly

Communication & Documentation

- Problem conditions documented (photos + notes)
- GC/owner informed of major substrate issues before installation
- Any "install over it anyway" decisions clearly understood and, if possible, written down

If you can honestly say **yes** to most of that list, you've given your carpet a fair shot at doing its job.

And if someone calls you back later, you'll be able to look them in the eye and say:

"We prepped this floor right. Let's figure out what *else* is going on."

In the next chapter, we'll move from what's under the carpet to the thing everyone sees (or should never see):

Seams, patterns, and direction.

That's where all your measurement and prep choices show up in the final picture.

Chapter 6 – Seams, Patterns, and Direction

If carpet was magically twelve miles wide and came in perfect room-sized pieces, this chapter wouldn't exist.

But it isn't. In the real world:

- Rolls are limited
- Rooms are weird
- Corridors are long
- And somewhere, you must join pieces together

That joint is your seam.

- A structural joint
- A visual joint
- A test of your prep, planning, and patience

If you treat seams like "just stick the two edges together," you'll get:

- Peaking
- Shadow lines
- Frayed edges
- And callbacks

If you treat seams like **surgical work**, you'll get:

- Joints that hold
- Lines that disappear in regular use
- And floors that feel like one piece even when they're not

This chapter is about turning seams from scars into signatures.

"Invisible" vs. "Quiet" Seams

LET'S GET SOMETHING straight:

There's no such thing as a truly invisible seam in every light, at every angle, forever.

Under the right combination of low-angle light, carpet style, and human obsession.

Someone can find it.

The goal isn't magic. The goal is:

- **Structurally solid** – it holds up under traffic
- **Visually quiet** – it doesn't jump out at normal people doing normal things
- **Compatible with the room** – it works with pattern and direction instead of fighting them

"Quiet" seams, done right, disappear into the story of the room. They're there, but nobody's having meetings about them.

Seam Placement: Where, Not Just How

YOU ALREADY DID A LOT of this work in your planning chapter, but let's hit it here from the seam's point of view.

When deciding where to put seams, think in layers:

1. Traffic

ASK:

- Where do feet *actually* go?
- Where do carts, wheelchairs, chairs, or luggage roll?

Guidelines:

- Avoid seams dead center in the main path if you can
- Tuck seams closer to walls, under furniture lines, or in low-traffic zones
- In corridors, consider running goods lengthwise to reduce cross seams (when practical)

Sometimes you have no choice. When a seam *must* cross traffic:

- Make it structurally perfect
- Accept that it might show a little but won't fail

2. Light

WHERE DO YOU HAVE:

- Strong sunlight from windows or doors?
- Rows of overhead cans or wall sconces?

Peaking and subtle height differences throw shadows.

Whenever possible:

- Run seams **parallel** to strong light sources, not straight across them
- Avoid placing seams where low-angle light will rake right across and highlight them

You can't control the sun. You *can* control whether your seam sits in its spotlight.

3. Subfloor & Joints

WE JUST SPENT A WHOLE chapter on prep. Now apply it:

- Don't sit seams directly over:
    - Big cracks
    - Active control joints
    - Hollow spots
    - Bad patches

If you have no choice:

- Make sure the subfloor is as solid and flat as you can get it
- Note it when you talk to the GC/owner

Seams want solid, flat ground—just like your knees do.

Seam Preparation: The Cut Is Half the Battle

YOU CANNOT FIX A BAD cut with an iron.

Once the face yarn is hacked or the backing is chewed, you're just dressing up damage. Seam prep starts with:

1. Cutting on a Row (When Possible)

FOR MOST TUFTED CARPETS:

- Locate a row in the face yarn
- Spread it slightly with your fingers or a tool
- Cut carefully along that row from the back or face (depending on your method)

Row cutting:

- Reduces random yarn loss
- Gives you cleaner, more predictable edges
- Makes the seam line blend better

Random "freehand" cuts across yarns?

- More fuzz
- More fray
- More obvious line

Use the carpet's construction to your advantage.

2. Straight, Consistent Edges

YOUR TWO SEAM EDGES must:

- Be straight

- Match each other
- Not serpentine in and out

Common problems:

- Cutting one piece straight and the other with little waves
- Cutting too much off one side and trying to "stretch it back"

Take your time.

Always better to:

- Cut a factory edge rather than trust it blindly
- Make sure both edges have been cleaned up and trued before you bother with tape

3. Seam Sealing (Don't Skip)

SEAM SEALING IS BORING—UNTIL you're staring at a split seam six months later.

For most residential and many commercial stretch-in jobs:

- Apply a bead of seam sealer (latex or manufacturer-approved sealer) to the **backing edge**
- Light bead—enough to bind yarn and backing, not enough to soak to the face
- Let it set to the recommended tackiness before heating, or follow product guidelines

On direct glue:

- Many manufacturers require **sealing all cut edges**, even at walls and transitions
- Do it. Loose yarn and delamination at edges looks terrible and is hard to fix cleanly

Seam sealer is cheap insurance.

Seaming Methods: Different Jobs, Different Rules

WE WON'T GO INTO EVERY specialty system on Earth, but here are the basics you'll live by.

1. Stretch-In Over Pad

THIS IS YOUR CLASSIC residential and some commercial work.

Key points:

- Seam tape chosen for:
    - Carpet backing
    - Traffic level
    - Installation method (over pad vs. direct)
- Iron temperature:
    - Hot enough to melt adhesive and wet the backing
    - Not so hot that you scorch or burn the carpet

Process (simplified):

1. Tape centered under seam, shiny/glue side up
2. Iron runs at a steady pace, heating glue as you go
3. Edges brought together gently—no smashing or over-compression
4. Seam rolled and weighted as it cools
5. Allow set time before stretching to full tension

Stretch direction:

- Ideally, stretch **across** seams, not along them
- Over-stretching parallel to a fresh seam can encourage peaking

2. Direct Glue-Down

NO PAD, CARPET GLUED straight to the subfloor.

Key differences;

- No seaming iron
- Seams made in dry laid carpet and then glued, or made into wet adhesive (depends on system and product)
- Sealed with seam sealer
- Correct notch trowel and consistent adhesive spread are critical

General ideas:

- Avoid big ridges of adhesive directly under seams
- Roll seams thoroughly to set them into the glue bed
- Respect open time and working time for adhesive—you don't want a skin that blocks proper bond

3. Double-Stick / Cushion-Backed Systems

CARPET GLUED TO PAD, pad glued to floor, or cushioned back carpet glued directly.

Key points:

- Manufacturer instructions matter a lot here—different backings need different adhesives and techniques
- Seam tape required different (sometimes wider or specialty types)
- Compression and cooling must be controlled to avoid diving, ridging, or telegraphing

Whatever the system is, don't wing it.

Read the data sheet, or call someone who has done that system before.

Tape, Iron, and Hands: The Mechanics

ONCE YOUR EDGES ARE prepped and your method chosen, the actual act of making the seam is like cooking:

- Right ingredients
- Right heat
- Right timing

Tape Placement

- Center the tape under the seam path
- Keep it straight—no hard kinks or curves
- Make sure the tape isn't riding up on debris, ridges, or fasteners

IN SOME SYSTEMS, YOU might use:

- Wider tape
- Different tape for cushion-back or high traffic

But the basics are the same: a smooth, even glue surface under both edges.

Iron Speed and Heat

EVERY IRON HAS A SWEET spot. So does every tape.

General habits:

- Don't run the iron so fast the adhesive doesn't fully melt
- Don't leave it so long in one spot that you burn or overcook the adhesive
- Keep a consistent pace once you find the right melt "feel"

You're aiming for:

- Wet enough to bond
- Not so flooded that it creeps everywhere or soaks through

Marrying the Edges

AS YOU PULL THE IRON along:

- Bring the two edges together gently with your hands or tools
- Don't mash them—just snug them together without crushing the face yarn
- Make sure pattern alignment is held where needed (more on that in a moment)

Watch for:

- Overlap (one side riding on top of the other)
- Gapping (tiny split that will look huge later)

Take the time to correct as the seam is warm and workable.

Rolling and Weighting

ONCE THE SEAM IS CLOSED:

- Use a seam roller to blend the face fibers over the joint
- Apply seam boards or weights to hold the seam flat while it cools
- Don't drag heavy metal straight across the face where it'll leave scars

Cooling time:

- Let the seam set before you stretch aggressively or open the area to traffic
- The adhesive needs time to firm up and hold its shape

Patience now is cheaper than repairs later.

Patterns: Taming Bow, Skew, and Drift

PATTERNED CARPET IS where seam skills really show up.

Patterns will:

- Bow (arc across the field)
- Skew (one side ahead of the other)
- Shrink or relax differently after unrolling

Your job is to manage all that as best you can.

Checking Pattern Before You Start

UNROLL AND CHECK:

- Does the pattern line cross the width in a straight line, or is it bowed?
- Are motifs lining up at consistent intervals, or drifting?

If it's wild:

- Talk to the rep or supplier before cutting up the whole roll
- You might be dealing with something defective, or you may need more time/material budget for pattern match

Aligning at the Seam

AT THE SEAM:

- Pick key motifs to align (centers, borders, or repeated shapes)
- Match them at eye-level view, not just somewhere nobody will ever look
- Use pins, weights, or other methods to hold alignment while seaming

Don't chase pattern "perfect" over a whole field if it destroys the layout elsewhere. Decide:

- Where must it look right? (main view, entry, centerline)
- Where can you let minor drift hide? (under furniture, near side walls, around bends)

Using Stretch to Your Advantage

SOMETIMES YOU CAN:

- Stretch a little more on one side to "walk" the pattern back into alignment
- Ease the pattern in small increments instead of taking all the correction in one seam

Key word: **small.**

Over-stretching to fix pattern can:

- Distort the backing
- Create tension problems
- Set you up for future wrinkles

Use correction stretches like a scalpel, not a hammer.

Direction: Pile, Nap, and How the Floor Feels

CARPET HAS DIRECTION even when it looks plain.

Pile (nap) lays a certain way. That affects:

- Color perception
- Sheen and shading
- The way footprints and vacuum marks show

Keeping Direction Consistent

- All pieces in a room or continuous area should run the **same direction**
- All seams should respect the manufacturer's guidelines for pile direction

IF YOU FLIP A PIECE 180°:

- It may look darker or lighter
- It may show a visible "change" line even with a perfect seam

Sometimes specs allow turning goods (e.g., in some commercial carpets), but don't assume. When in doubt: same direction.

Direction vs. Light and Traffic

SOME GENERAL PREFERENCES:

- Running pile **toward** the primary light source can minimize shading on some styles
- Running pile **with** the main traffic direction can improve appearance in some corridors

But different styles behave differently. Trust:

- Manufacturer recommendations
- Your own experience with that product type

Don't let "direction games" ruin seam quality. A solid seam with acceptable shading is better than perfect theoretical direction and a terrible joint.

Common Seam Problems (and How to Avoid Them)

LET'S BREAK DOWN THE big four.

1. Peaking

THE SEAM LOOKS LIKE a little ridge or tent.

Causes:

- Over-tensioning carpet **across** the seam without proper cooling/ pattern
- Seams too tight, edges forced together with too much pressure
- Tape too stiff or too narrow for the situation
- Subfloor hump directly under the seam

Prevention:

- Use proper stretch sequence and let seams cool/set before heavy stretching
- Don't crush the edges together—just meet them
- Pick correct tape and prep the subfloor flat under the seam path

2. Gapping

VISIBLE LINE WHERE the backing is visible between edges.

Causes:

- Uneven trimmed edges pulled apart under tension
- Poorly aligned pieces during seaming

Prevention:

- Trim accurately, no jagged or undercut areas
- Hold edges snug during seaming—no "we'll pull that together later"

thinking
- Consider environmental factors (big temp swings, humidity)

Minor gaps can sometimes be disguised. Big ones are usually there for life.

3. Fraying and Fuzzing

YARN LOOSE ALONG THE seam, fuzz ball build-up, ragged line.

Causes:

- No seam sealer
- Over-aggressive cutting through face yarns

Prevention:

- Seal cut edges when required/recommended
- Row cut when possible
- Use soft rollers, gentle tools on the face

4. Telegraphing and Tape Lines

VISIBLE LINE WHERE the tape or joint reads through thin carpet.

Causes:

- Thick tape or ridges right under thin carpet
- Direct glue installs with ridged adhesive under seams
- Subfloor uneven under the seam

Prevention:

- Use appropriate tape for carpet thickness and installation type
- Flush, even adhesive bed under seams
- Level subfloor under seam lines

If you can feel the seam ridge with your hand before carpet goes down, expect to see it with your eyes after.

Job Note – The Seam That Taught Me to Slow Down

I DID A CORRIDOR ONCE with a patterned commercial carpet—tight weave, strong lines, tons of light blasting down the run. The schedule was already wrecked before we started: material showed up late, guests were still in the building, and the GC wanted the hallway open "yesterday." On paper we were fine. We'd prepped alright, measured alright... and then we tried to make up all that lost time at the seam.

I ran the iron a little hotter and a little faster than I should have. We pulled the edges together just a bit too tight and didn't give the seam much time to cool before we started stretching. On install day, it looked good enough. Two weeks later, in the afternoon sun, every seam in that corridor showed a slight ridge. Not a disaster—but enough that people walking on it noticed, the GC noticed, and I saw it every time I closed my eyes.

We went back, eased some tension, and cleaned up the worst spots, but the real problem wasn't skill—it was speed. Those seams didn't fail because I didn't know what I was doing. They failed because I didn't give the work the time it needed.

Now, when a schedule is insane and somebody says, "Just stretch it, we need this open," I remember that hallway. Sometimes the bravest, most professional thing you can say is, "This seam needs thirty minutes to cool, or you're going to be calling me back."

Checklist – Seam, Pattern, and Direction Basics

USE THIS AS A QUICK mental check before you start running iron.

Placement

- Seams are placed with traffic in mind (not all dead center in the main path)
- Seams consider light direction where possible (not across the worst low-angle sun)
- Seams avoid major cracks, joints, and bad patches when possible

Preparation

- Edges cut cleanly, ideally on a row where appropriate
- Edges are straight and matched, no waves
- Seam edges are sealed when required/recommended
- Tape type is appropriate for carpet and installation method

Execution

- Tape is centered, flat, and not sitting on debris or ridges
- Iron heat and speed dialed in before full run
- Edges brought together snugly but not forced or crushed
- Seam rolled and weighted properly while cooling
- Adequate cooling time allowed before full stretch or traffic

Patterns

- Pattern repeat checked for bow/skew before cutting
- Key motifs aligned at seams, especially in main view areas
- Any cheats or compromises planned where they'll be least noticeable

Direction

- Pile/nap direction consistent across connected areas

- Direction chosen with manufacturer guidance and appearance in mind
- No random flipped pieces just to "save material" unless explicitly allowed and acceptable

If you can say **yes** to most of that list, you're not just "making seams."

You're building joints that respect the room, the material, and your own name.

Chapter 7 – Stretching and Tension

Wrinkles don't just "show up."

They're decisions you made a year or two earlier:

- To skip the stretcher because "it's just a small room"
- To kick a whole floor because "that's how we've always done it"
- To ignore manufacturer requirements because "I've never had a problem"

Then one day, you get the call:

- "The carpet's buckling in the hallway."
- "We've got waves in the conference room."
- "The hotel corridor looks like a washboard."

And suddenly, everyone's staring at the floor—and at you.

This chapter is about **tension**:

- What it really does
- Why a power stretcher isn't optional
- How to stretch rooms and corridors so they stay flat
- And when you can use a knee kicker without lying to yourself

You can cheat seams and prep sometimes and get lucky.

You can't cheat tension forever.

Why Tension Matters

CARPET IS NOT A PAINTING you hang on a wall. It's a flexible surface that:

- Expands and contracts with temperature and humidity
- Gets pushed and pulled by traffic and furniture
- Slowly relaxes from whatever tension you set it at

If you don't stretch it enough:

- Traffic will work the slack out into wrinkles
- Humidity swings will puff up low-tension areas
- Seams may take the movement and fail

If you stretch it properly:

- The carpet fights back *against* movement instead of giving in
- Wrinkles are far less likely to form
- The floor feels solid underfoot—no "squishy" or loose feel

Think of it like a drum:

- Tight drum = clean sound, responsive surface
- Loose drum = dead sound, saggy skin

You're tuning the floor.

Knee Kicker vs. Power Stretcher: What They're For

LET'S SETTLE THIS ONCE.

What a Knee Kicker Is For

A KNEE KICKER IS A **positioning tool**, not a main stretching tool.

Good uses:

- Hooking carpet onto strip along the edges once the main stretch is set
- Working in tight areas (closets, small bathrooms)
- Making small adjustments at corners and around obstacles
- Bumping carpet into tackstrip in stairs and tricky spots

Bad use:

- "We stretched the whole room with the kicker."

That's not stretching. That's "shoving and hoping."

What a Power Stretcher Is For

A POWER STRETCHER IS designed to:

- Apply consistent, controllable tension across the room
- Let you stretch from wall to wall, not half-kick at random
- Meet or get closer to manufacturer tension requirements

A stretcher with:

- A solid head
- Enough poles for the room
- A stable tail block

...is the closest thing you have to a guarantee that the carpet is actually tight, not just "looks okay today."

If you're installing stretch-in carpet in regular-sized rooms and not using a stretcher, you're not just cutting corners—you're setting up future callbacks.

Tension Targets (Without Turning You Into a Spec Robot)

MANUFACTURERS OFTEN specify things like:

- A certain percentage of stretch
- Tension levels per foot
- Detailed power stretcher setup requirements

You might not walk around quoting those numbers, but you should understand the idea:

- Carpet needs to be **pulled** beyond its relaxed length
- That tension then gets "locked in" by tackstrip and walls
- When the carpet tries to relax, it pulls against the pins and stays flat

Street-level rules:

- If you're just "snugging" the carpet onto the strip, it's not stretched
- If you're afraid to lean into the stretcher at all, it's not stretched
- If you can grab a handful of carpet in the middle of the room and easily pull a small wave, it's probably under-stretched

You don't want to warp the backing or over-tension to the point of damage. You want:

Firm, even tension across the space, not just "no obvious slack."

Stretching a Standard Room: The Basics

EVERY INSTALLER HAS their own dance, but the principles stay the same.

1. Setup the Room

BEFORE STRETCHING:

- Carpet is cut to size with some excess for trim
- Seams are made, cooled, and set (if present)
- Tackstrip is properly installed and secure
- Subfloor is prepped and reasonably flat

Remove:

- Major obstacles
- Furniture (or work in pre-planned phases)
- Anything that blocks you from placing the stretcher head and tail

You can't stretch around a room packed with junk and expect good results.

2. Place the Tail Block

THE TAIL BLOCK NEEDS:

- A solid, clean surface to push against (wall or jamb)
- Protection for walls if needed (pads, blocks, scrap)

Make sure:

- The tail isn't sitting on loose debris or weak baseboard
- It won't slip or punch through drywall when you load the stretcher

If you've ever seen a stretcher blow a hole in a wall, you don't forget it.

3. Set the Head and Poles

POSITION:

- Stretcher head on the opposite side from the tail block
- Teeth set into the carpet at a proper angle—not just barely catching the backing

Poles:

- Straight line from head to tail
- Joints firmly locked
- Enough poles to comfortably reach, not barely stretched out

If poles are bowing like fishing rods, something's off:

- Either you're cranking too hard, or
- The poles setup is wrong for the room

4. Stretch Sequence

A COMMON PATTERN FOR a rectangular room:

1. Anchor one short wall about 3 ft. (rub and hook with stair-tool)
2. Stretch toward the opposite short wall in 3 ft. width:
    a. Set
    b. Apply tension
    c. Hook carpet onto tackstrip
    d. Release stretcher
    e. rub and hook with stair-tool
3. Once that wall is hooked:

a. Set wall by rubbing down hooking with stair-tool

1. Repeat steps 1-3 on long wall (now you have your set walls)
2. move the stretcher so you're working from the set wall toward the next long side, working around the room
3. Finish in the upset corner that you started in

5. Trim and Tuck After Stretch

DON'T TRIM EDGES TOO tight **before** you've hit final tension. Stretch first, then:

- Trim to a clean, consistent gully
- Tuck neatly with stair tool or bolster
- Check that you haven't left baggy corners

If you trim too tight before final stretch, you'll either:

- Starve the edges (no tuck room), or
- Cut so short you see backing at the baseboard

Neither is a good look.

Corridors and Large Areas

ROOMS ARE ONE THING. Long hallways and big open spaces are another.

Corridors

CHALLENGES:

- Long runs where tension can "stack up"
- Seams along or across the corridor
- Fixed conditions at doorways and walls

Habits:

- Use adequate stretcher extensions or mid-room tail blocks as needed
- Stretch in **sections**, not just one big pull from one end
- Anchor key areas (like near elevator or main entry) before working down the run

Watch for:

- Wrinkles forming mid-corridor after you fix one end
- Seams that try to peak when you pull too hard across them

You may need:

- Multiple stretch directions
- To use different starting points depending on the building layout

Large Open Areas

BALLROOMS, BIG OFFICES, multipurpose spaces:

- Rarely perfectly square
- May have columns, stage areas, or strange features

Approach:

- Decide your **primary stretch direction** first
- Work outwards from a solid anchor line (a main wall, a central seam, or a key pattern alignment)
- Use multiple stretcher placements to pull evenly across the whole field

Don't:

- Stretch randomly from different sides with no plan
- You'll end up with tension fighting itself or "dead zones" with slack trapped in the middle

Re-Stretching: When the Past Comes Back

SOMETIMES YOU WALK into a job where the carpet's already waving at you.

Re-stretching is:

- Often better than replacing
- Sometimes a band-aid over deeper issues
- Almost always a physical grind

When Re-Stretching Makes Sense

GOOD CANDIDATES:

- Carpet in decent shape
- Issues mostly appear as wrinkles/buckles
- Subfloor is solid enough
- Tackstrip still has bite

Situations where re-stretching is more questionable:

- Backing shot or delaminating
- Severe water damage or heavy contamination
- Junk tackstrip and loose substrate everywhere

You can re-stretch a bad install. You can't re-stretch bad material back into new.

Process Basics

YOU'LL GENERALLY:

1. Move or remove furniture as needed
2. Release edges and sometimes seams in the affected area
3. Use a power stretcher to pull carpet tight again

4. Re-trim and re-tuck at walls
5. Re-make or re-tune seams if they've been stressed or damaged

It's often harder than doing it right the first time. But sometimes it's the best option for the client.

Checking Your Work: Is It Really Tight?

HOW DO YOU KNOW IF you've stretched enough?

- Trust your "feel" after enough experience
- Use manufacturer field tests (like trying to pinch and pull a wrinkle up)

Street tests:

- Try to grab a wrinkle in the center of the room with your hand
    - If you can easily pull up a ridge, that's not great
- Walk the room in multiple directions
    - Feel for loose or hollow zones underfoot
- Watch how the carpet reacts:
    - Do you see small "rivers" moving toward you as you walk? That's slack.

The goal is not to have the carpet banjo-tight and stressed to death. The goal is:

Firm, even resistance across the space.

If you're unsure, it's usually safer to give it **one more controlled stretch** than to stop short and hope.

Common Stretching Mistakes

LET'S HIT A FEW YOU'LL see (or make) sooner or later.

1. "I Can Kick This, It's Small"

SMALL ROOMS STILL MOVE.

Even in a 10' x 10' bedroom:

- Under-tensioned carpet will relax and wrinkle eventually

- Heat and humidity swing will work looseness into the field
- That cheap, fast "kick-only" job becomes a re-stretch later

You **can** use a kicker more heavily in:

- Tiny closets
- Tight little spaces where a stretcher simply won't fit

But "small" doesn't automatically mean "no stretcher."

2. Stretching Before Seams Are Ready

IF YOU STRETCH WHILE the seam is still warm or hasn't fully set, you can:

- Distort the seam
- Create peaking
- Break the bond before it even fully forms

Give seams:

- Proper cooling and set time
- Especially in critical visual areas

Speed can cost you more time later.

3. Ignoring Subfloor Issues

IF THE SUBFLOOR:

- Has high ridges
- Has low soft areas
- Has bad patch right under where you apply tension

Your stretch may:

- "Bridge" over voids
- Telegraphed problems through faster
- Make wrinkles appear where the floor dips or flexes

Stretching doesn't fix bad prep. It just stretches carpet over it.

4. Uneven Stretching

IF YOU:

- Over-stretch one area
- Barely stretch another
- Or stretch from multiple sides with no plan

You can trap slack in:

- Corners
- Mid-room zones
- Around columns

You might not see it on day one, but those zones tend to wrinkle first.

Plan your stretch path. Work systematically.

Protecting the Carpet While You Stretch

TENSION IS ONLY HALF the equation. You also must keep from **hurting** the carpet while you apply it.

Habits:

- Make sure stretcher teeth are sharp, but not tearing yarns
- Don't let poles chew into delicate trim or drywall—use pads
- Avoid dragging the stretcher head across the face while loaded
- Be careful at thin areas or where backing is already stressed

You want the carpet to remember the tension, not the damage.

Job Note – The Hotel That Called Me Back… Twice

THERE WAS A HOTEL JOB where everything was stacked against schedules.

We were behind before we even started. Material delays, other trades in the way, guests in the building. We were working nights and trying to keep as much of the corridor open as possible so the front desk wouldn't revolt. On paper, we were doing it right: good pad, proper tackstrip, power stretcher on site.

In reality, we were rushing the sequence.

We stretched, but we cut corners on the resets. We tried to "make up time" by not double-checking every pull. We convinced ourselves that "good enough for tonight" was the same as "good enough for the life of the carpet."

A few months later, the hotel called.

"We're seeing some slight bubbles in the corridor."

We went back and found a couple of zones where we'd clearly not pulled as hard. We tightened them up, stretched things smoother, felt better about it, and left.

Six months after that, they called again.

"It's doing it again, further down."

This time we walked the whole run slowly, from one end to the other. Every place that was starting to wrinkle was part of the area we'd worked on late in the shift—right when everyone had been tired and in a hurry. The stretcher had been there the whole time. The carpet was fine. The subfloor was fine.

The weak link was us.

That job taught me that owning a power stretcher doesn't mean you've done your job. You must use it like you mean it, every time, especially when you're worn out and just want to get the last section done and go home.

These days, when I feel that "let's just knock this out quick" voice in my head toward the end of a long night, I hear that hotel's front desk calling about waves in the corridor. It keeps me honest.

Checklist – Stretching and Tension

USE THIS AS A QUICK mental rundown before and after you stretch.

Before Stretching

- Subfloor is prepped and reasonably flat
- Tackstrip is solid and spaced correctly
- Seams are made, cooled, and set (where applicable)
- Room or area is as clear as realistically possible
- I have enough stretcher poles and a stable tail block setup

During Stretching

- I have a clear stretch sequence (not random pulling from everywhere)
- I'm anchoring one side and working across the field logically
- I'm using the kicker for positioning and detail, not as my primary tension

After Stretching

- There are no obvious "pools" of slack in corners or along walls
- Edges are trimmed and tucked neatly after final tension, not before
- I can walk the area in multiple directions without feeling movement underfoot

If you can honestly check those boxes, you're ahead of most installers.

You're not just **covering** floors. You're **tuning** them so they stay flat for years.

Next up, we're going to take all that tension and precision and aim it at the spots that expose bad work faster than anything else:

Stairs, landings, and tricky details.

That's where your skill shows up in every single step someone takes.

Chapter 8 – Stairs, Landings, and Tricky Details

If you want to know how good an installer really is, don't look at the middle of the room.

Look at:

- The stairs
- The landings
- The little returns around a post or closet
- The place where the carpet meets tile, metal, or wood and has nowhere to hide

Anyone can make a big open field look decent from 10 feet away.

Stairs and details tell the truth at 10 inches.

Stairs are the job interview you can't fake.

This chapter is about making those high-exposure areas look intentional, not accidental—so that every step someone takes says, "Whoever did this knew what they were doing."

Why Stairs Expose You

STAIRS ARE BRUTAL BECAUSE:

- Every tread is a **test**: people step right on your work, day after day.
- Lighting often hits at angles that show every lump and wrinkle.
- The human eye expects rhythm—each step should look like the last one.

When stairs are done badly, you see:

- Uneven nosings
- Loose carpet on treads
- Wrinkles on risers
- Gaps at stringers
- Tacker staples everywhere the eye goes

When they're done right, you don't see "stair carpet."

You see clean rhythm, safe footing, and a smooth ride up and down.

Know Your Stair Types

YOU'LL SEE A FEW COMMON setups. How you approach them depends on what the designer/owner wants and what the building allows.

Straight Run Stairs

- Uniform treads and risers
- No curves, no winders
- Most common in homes, basic commercial, back-of-house areas

Open-Sided Stairs

- One or both sides open
- Exposed stringers, sometimes with decorative nosings
- Often in nicer homes, lobbies, or feature stairs

MORE VISIBILITY = MORE pressure to get it right.

Winder / Pie Stairs

- Steps that turn a corner as they climb
- Treads wedge-shaped (wider on one side than the other)

Box Stairs / Enclosed

- Walls on both sides
- No exposed stringer, nosing hidden under wall or skirt board

CLEANER VISUALLY, BUT still honest about your cuts and tucks.

Common Installation Styles

THERE ARE SEVERAL WAYS to dress stairs. You'll hear different terms in different regions, but these are the big three.

1. Waterfall

CARPET:

- Runs over the nosing
- Drops straight down the riser in one continuous piece
- No defined angle at nosing—just a soft fall

Pros:

- Faster
- Forgiving on some materials
- Common in basic residential and commercial

Cons:

- Less tailored appearance
- Can show more slack or bagging if poorly done
- Less crisp around nosing

2. Cap-and-Band (Bullnose / French Cap style)

- "Caps" the tread and nosing tightly
- Then a separate "band" piece covers the riser

PROS:

- Clean, tailored look
- Shows craftsmanship on each step
- Great for pattern control

Cons:

- More time
- More cuts and seams
- Easier to mess up if rushed

3. Hollywood / Tailored Wrap

- Pulled tight around the nosing
- Stapled or fastened close to the underside or return
- Often wrapped more tightly to the profile of the step

PROS:

- Very tight, custom look
- Popular for premium residential and some feature stairs

Cons:

- Unforgiving: mistakes show instantly
- Requires clean nosing and good substrate

Prep: Stairs Before Carpet

JUST LIKE FLOORS, STAIRS will tell on you if the structure is bad.

Check:

- Treads for **looseness or squeaks**
- Nosings for damage, rot, or broken edges
- Risers for cracks, gaps, or loose panels

Fix what you can:

- Screw down loose treads from above or below
- Repair damaged nosings with proper filler or replacement
- Fill big gaps where carpet might dip or collapse

Clean:

- Remove staples, nails, old pad, and debris
- Scrape old glue or padding residue
- Vacuum steps and stringers

If the stair itself is junk, your beautiful carpet is just lipstick on it. Do what you can to make the structure solid before dressing it.

Padding on Stairs

PAD ON STAIRS IS NOT just about comfort—it affects safety and appearance.

Basics:

- Use pad that's **approved** for stairs usually 3/8" or 7/16"—denser, not spongy thicker pad that lets carpet roll over itself.
- Cut pad **tight** to the step and secure it well:
    - Staples where appropriate
    - No loose pad flopping over nosing

Common approach:

- Staple on tread-back close to tac-strip and top riser
- Pad cut to fit snug between strips, not overlapping onto tac-strips and causing bumps

You want:

- A smooth cushion under the carpet
- No pad "lip" at nosing that creates a rollover or lump
- Solid anchoring so pad doesn't migrate over time

Layout: Making Every Step Look Like It Belongs

STAIRS ARE REPETITIVE. That's your enemy and your friend.

If one step looks off, all the others make it more obvious.

Measure Once, Use Often

PICK **standard dimensions** and stick to them: example 42.5" x 20" (x9)

- Width- Overhang / wrap distance
- Length- Tread depth (bottom to back wrap distance)

Consistency beats "eyeballing it" twelve times and hoping it comes out even.

Pattern Considerations

WITH PATTERNED CARPET:

- Decide where you want the pattern to "hit":
    - At the nosing?
    - At the riser midpoint?
    - Following through landings and corridors?

Try to:

- Keep a consistent pattern element on each tread (e.g., a line near the nose)
- Avoid random pattern jumps from step to step unless design demands it

Waterfall Basics: Clean and Fast (Not Sloppy)

FOR A WATERFALL INSTALLATION on straight box stairs:

1. **Cut Length & Width**
    a. **Measure from top of top riser to bottom of bottom riser**
    b. **Measure each step width. If square top to bottom, row cut it**
    c. ***If the stairs are not square, do one or two at a time***
2. **Anchor Starting Point**
    a. **Usually start at the bottom**
    b. **Hook carpet securely by rubbing the pins into the carpet with a hammer**
3. **Form Each Step**
    a. **Pull carpet over the nose**
    b. **Smooth it straight down the tread with your hand or stair tool**
    c. **Use a kicker to bump the carpet starting in the center and working towards each side**
    d. **Tap stair tool with rubber mallet to snug into gully at tread/riser evenly from the center out**
4. **Trim Cleanly**
    a. **Trim with wall trimmer and tuck with stair-tool**
5. **Check Rhythm**
    a. **Stand back and look:**
        i. Are riser faces smooth?
        ii. Are nosings consistent?
        iii. Any bagging or sagging?

Cap-and-Band / Tailored Steps

THIS IS WHERE YOUR precision gets tested.

General approach (one common method):

1. **Cap (Tread) Piece**
    a. **Cut piece to cover tread + nosing + a bit under or down the riser front**
    b. **Dry-fit to make sure the nosing wraps crisp and even**
2. **Wrap the Nosing**
    a. **Pull carpet tightly over the nose**
    b. **Staple or fasten underneath or on the hidden part of the nose**
    c. **Form a sharp angle, not a bulging roll**
3. **Band (Riser) Piece**
    a. **Cut a separate piece for the riser**
    b. **Fit it from tread to tread (or under the cap depending on method)**
    c. **Staple or fasten along the edges, keeping the face smooth**
4. **Repeat Consistently**
    a. **Use your story stick so each cap and band is the same size**
    b. **Keep side lines straight along stringers**

The visual goal:

- Crisp nosings
- Smooth riser faces
- No random bulges or dips

It takes longer. But when done right, every step looks like it was drawn with a ruler.

Open-Sided Stairs and Stringers

OPEN-SIDED STAIRS EXPOSE your work along the edges.

Key points:

- Cut edges must be **sealed** where required and handled with care
- Tucks along open stringers must be consistent and clean
- Overhangs should be symmetrical, not drifting in and out

Approach:

- Often, you'll wrap carpet slightly over the edge and under a lip or trim
- Tucks should sit tight into the stringer junction—no gaps, no torn edges
- Staples or fasteners should be:
  - Close to the edge
  - Consistent
  - Hidden as much as the structure allows

If you're working with a painted or stained stringer:

- Protect the finish
- Don't shred it with tools
- Use padding or tape where tools ride the surface

Winders and Pie Stairs: The Brain Twisters

WINDERS ARE THE STEPS that turn a corner as they go up.

Challenges:

- Treads wider on one side than the other
- Pattern distortion
- Keeping the eye from catching weird angles

Basic strategy:

1. **Find the Focal View**
    a. **Will people normally see the stairs mostly from the bottom looking up?**
    b. **Or from the top looking down?**
    c. **Choose one view to prioritize.**
2. **Align Key Edge**
    a. **Often, you'll keep the outer (wide) edge visually consistent, as that side usually draws the eye**
    b. **The inside corner can handle a bit more cheat because it's tight and shadowed**
3. **Template Your Pieces**
    a. **Use cardboard or scrap to template tricky shapes before cutting the "real" material**
    b. **Mark grain/pile direction clearly on your templates**
4. **Commit and Blend**
    a. **Once pieces are cut, install with attention to flow:**
        i. Do lines feel like they "turn the corner"?

Landings: Mini Rooms That Matter

LANDINGS ACT LIKE PAUSES in the stair run—and little rooms of their own.

They need:

- Proper subfloor prep like any floor
- Logical seam placement (if any)
- Directional planning so the flow up/down stairs looks natural

Consider:

- Whether to run carpet on landings the same direction as the stairs or as the next floor
- Where you'll place seams if landings are larger (near walls vs. in the

walking path)

If a landing feeds into a hallway:

- Make sure the transition feels intentional
- Avoid random direction changes that make the landing look like an afterthought

Transitions and Tricky Edges

STAIRS OFTEN TIE INTO:

- Tile
- Wood
- LVT
- Metal nosings or trims

Each of those can make you look sharp—or sloppy.

Key habits:

- Check heights on both surfaces
- Choose appropriate metal or transition profile
- Make sure carpet:
    - Is fully supported underneath at the edge
    - Is not hanging in mid-air over a lip
    - Is cut clean and, if needed, sealed at the edge

At the top and bottom of stairs:

- Make transitions flush where possible
- Avoid creating small trip lips
- Align carpet direction and pattern with the connecting floor if you can

Safety: It Has to Look Good AND Be Safe

PRETTY STAIRS THAT are unsafe are a lawsuit waiting to happen.

Watch for:

- Loose areas on treads that can grab toes or heels
- Bulging nosings that can catch feet
- Wrinkles on risers that telegraph into shadow lines and trip

perception

- Metal nosings not properly fastened

Also:

- Be careful with patterns that create optical illusions on stairs (zigzags, aggressive stripes)
- If the design calls for something wild, make sure the owner understands how it'll look in motion

Your work should:

- Feel secure underfoot
- Not pull eyes into weird depth illusions
- Give confidence, not hesitation, with every step

Job Note – The Condo Board That Almost Fired Me

I DID A HALLWAY JOB in a condo building once—occupied, lots of older residents, and a board that treated every little detail like it was a federal case. It was the kind of place where every decision went through a committee and every complaint came with a photo.

We had a solid plan going in. The work was phased floor by floor, notices went on doors, and we agreed on daytime hours only. On the install side, we were doing fine: prep was good, seams were clean, stretch was tight. The problem wasn't the carpet. It was everything around it.

On that first day, we tracked a little dust into the elevator, left a couple pieces of cardboard near the trash room "just for now," and leaned some tools in the corridor between trips. Nothing wild. Nothing we hadn't seen on a hundred other jobs. But in a building full of people walking those halls all day, it looked sloppy.

By the afternoon, one resident had taken pictures of everything and emailed the board with the subject line: "These guys are a mess." That email made it to the property manager, who pulled me aside and said, "If this is how it's going to go, we're going to stop the job."

The hall we'd finished looked good. The carpet itself was solid. But the way we were living in their common space—dust, scraps, tools parked where people walked—made it hard for anyone to see the quality of the install. We weren't just being judged on our seams; we were being judged on our footprint.

We regrouped fast. One guy on the crew became the dedicated "clean runner." We laid extra protection in the elevator and lobby. At the end of every day, we swept, vacuumed, and walked the whole corridor before we left. Nothing stayed in the common areas unless we were actively using it.

Two days later, the same resident who complained told the board we'd "really cleaned up our act" and were being very careful. Nothing about our install technique had changed. The difference was how seriously we took their space.

That job taught me that in occupied buildings, a clean, controlled jobsite is part of the install. The carpet can be perfect, but if people have to step over your mess to see it, all they'll remember is the mess

Checklist – Stairs, Landings, and Details

USE THIS CHECKLIST before you call a stair job finished.

Structure & Prep

- Treads are solid (no squeaks or movement)
- Nosings intact and not crumbling or broken
- Risers reasonably smooth (no huge gaps or loose panels)
- Old staples, nails, and debris removed
- Steps and stringers vacuumed before pad/carpet

Pad & Strip

- Pad suitable for stair use (not mushy junk)
- Pad cut to fit snugly, not lapping over nosings
- Tackstrip installed where needed, secure, and aligned

Layout & Cutting

- Chosen installation style (waterfall, cap-and-band, Hollywood) is clear
- Dimensions consistent from step to step (use story stick if possible)
- Pattern plan decided, especially for winders and landings
- Templates used on tricky or open-sided shapes when needed

Execution

- Treads are tight, no loose feel underfoot
- Nosings are clean and consistent (no sagging or bulging rollovers)
- Risers are smooth, no wrinkles or bagging
- Staples (if used) placed where they're as hidden as possible
- Open-stringer edges are cleanly tucked and consistent

Landings & Transitions

- Landings are prepped like any floor (flat, clean, solid)
- Carpet direction on landings and connecting floors feels intentional
- Transitions at top and bottom of stairs are safe and smooth
- Metals and trims are securely fastened, not rocking or sharp

Final Look & Safety

- Walked the stairs up and down multiple times—no catches or slips
- Viewed from top and bottom—line and pattern flow looks right
- No obvious trip hazards or weird bulges along nosings or edges

If you can honestly check those boxes, you've done more than "just put carpet on stairs."

You've built:

- A safe path
- A clean visual rhythm
- And a little piece of craftsmanship most people will never fully notice—but feel every time they use it.

Part III – Real-World Jobs

Chapter 9 – Working in Occupied Spaces

It's one thing to install in an empty house or building.

Nobody is there.

You can spread out your tools.

You can stash rolls wherever you want.

You can make noise, dust, and mess—and clean it up before anyone ever sees it.

It's another thing entirely to install in a place that's still alive:

In those spaces, you're not just an installer.

Your skill with people, setup, and cleanup matters as much as your skill with a knife.

This chapter is about how to work in occupied spaces without becoming the story.

Rule #1: You're in Their World, Not Yours

ON AN EMPTY JOB, THE site feels like yours.

In an occupied space:

- You're walking into someone's **business**, **home**, or **life**
- Your schedule might be their "off-hours"
- Your workspace might be their kitchen, bedroom, lobby, or main hallway

If you act like you own the place, you're going to clash with:

- Staff
- Homeowners
- Tenants
- Management

Start with this mindset:

"I'm here to do a great job **and** to make this as easy on them as possible."

That doesn't mean you become a doormat. It means:

- You communicate clearly
- You respect boundaries
- You act like the pro they hired

Communication: Before You Ever Unroll a Drop Cloth

MOST CHAOS IN OCCUPIED spaces comes from people not knowing what's happening, when, and for how long. You can't control everything, but you can control what you **say**.

Ask the Right Questions Up Front

BEFORE THE JOB STARTS, you should know:

- **Hours you're allowed to work**
    - Nights only? Weekend only? Certain quiet times?
- **Noise restrictions**
    - No demo during work hours? No grinders before 9 a.m.?
- **Access**
    - Which doors to use? Freight elevators only? Where to park?
- **Protection expectations**
    - Are they expecting plastic walls? Masonite? Floor protection over existing surfaces?
- **Where you can stage**
    - Which rooms or corners can hold tools and materials without upsetting anyone?

The more you know, the fewer arguments you'll have later.

Set Expectations with the Client

YOU DON'T HAVE TO GIVE them a novel, but you should say things like:

- "We'll need this area fully cleared before we start."
- "There will be some noise during demo and prep."
- "We'll need a path from the door to this room for moving material."
- "We'll clean up at the end of each day and leave a walkable path."

If they think:

- You're magic and can install without noise, dust, or disruption

...they're going to be upset no matter what.

If they understand:

- There will be some hassle, but you have a plan to control it

...you're already ahead.

Protection: Runners, Plastic, and Dust Control

YOU CAN DO PERFECT carpet work and still lose a client if you:

- Track dust through a space
- Scratch their floors
- Coat their desks in grinding dust
- Splash adhesive on their furniture

Protection is part of the job.

Floor and Path Protection

ANY TIME YOU'RE:

- Moving rolls through a lobby
- Rolling dollies over nice tile or wood
- Crossing finished floors in a home

...you should be using:

- Ram board / Masonite / similar for heavy paths
- Temporary runners or drop cloths
- Cardboard at pinch points (tight corners, elevator doors)

It's simple:

If your wheels, boots, or tools touch it and it's not yours, **protect it**.

Dust Control

SUBFLOOR PREP MAKES dust. That's not optional.

How much of it spreads is.

Habits:

- Use vacuums attached to grinders where possible
- Sweep and vacuum before you pack up for the day
- Close doors to rooms that don't need dust in them
- Use plastic sheeting or zip-walls when grinding or scraping near sensitive areas

If you're in:

- Medical spaces
- Food prep areas
- IT-heavy environments

...dust control is not just courtesy—it's sometimes a liability issue.

Protecting Furniture and Belongings

IN HOMES AND OFFICES:

- Cover nearby furniture with plastic or drop cloths if there's any chance of dust or spray
- Move items carefully and return them roughly where you found them
- Stay away from obviously sensitive equipment unless the owner moves it

If you break something:

- Own it
- Don't try to hide it
- Get the conversation going immediately

Your reputation is worth more than pretending that broken lamp never existed.

Working Around People: Noise, Behavior, and Boundaries

Noise: People remember how you made them feel more than how your seams look.

TOOLS MAKE NOISE. BUT you can control:

- **When** you're loud
- **How long** it lasts
- **Whether** you give any warning

Good habits:

- Save the loudest work (demo, grinding) for agreed-on windows
- Let nearby staff or residents know "We're going noisy for an hour" when possible
- Turn off radios or keep them at a low, professional level (and watch language on the music)

If you're in someone's house:

- Ask before playing music
- Keep language clean around kids
- Be the crew they're not afraid to have in front of their neighbors

Behavior

THIS SEEMS OBVIOUS, but:

- Don't argue loudly with your crew in front of clients
- Don't swear constantly
- Don't make comments about people walking by
- Don't park yourself in their break room like you own it

Also:

- Watch bathroom use—ask what you're allowed to use and keep it clean
- Don't leave food trash or drink cups all over their space
- Take smoke or vape breaks away from entrances and open windows

You're always being judged even when nobody says anything.

Boundaries

REMEMBER:

- In a house, you're in someone's personal world
- In an office or hotel, you're in someone's professional world

Stay out of:

- Drawers
- Cabinets
- Rooms that aren't part of your scope

Don't:

- Comment on their money, décor, or personal life
- Wander around "exploring" the building without permission

Professional means:

"I'm here to work. I go where I'm needed. Then I'm gone."

Hotels and Multi-Unit Properties

HOTELS, APARTMENTS, condos, senior living—anywhere with lots of people living or staying—have extra complications.

Hallways and Common Areas

YOU'RE WORKING IN THE main veins of the building.

Think about:

- **Access**: Can people still get to elevators, exits, and rooms safely?
- **Signage**: Use cones, tape, signs: "Wet Adhesive / Fresh Carpet – Please Use Other Side"
- **Phasing**: Work one side of the hallway at a time when possible to maintain a path

If you're working nights:

- Keep voices and impact noise controlled
- Coordinate with management about quiet hours and the noisiest tasks

Rooms

WHEN WORKING ROOM-BY-room:

- Make sure front desk or property manager knows which rooms are out of service
- Keep tools and trash in the rooms you're working, not all over the corridor
- Close doors when you're not actively going in and out

If guests or residents try to slip into your work zone:

- Stay calm
- Explain what's safe and what's not
- Help them navigate around if you can

You're not security, but you **are** responsible for not creating unsafe situations.

Offices and Commercial Spaces

OFFICES ARE FULL OF:

- Computers
- Wires
- Furniture
- People on deadlines

If you're not careful, you'll spend more time moving stuff than installing.

Furniture and Electronics

BEFORE YOU START:

- Clarify who moves what:
    - Are you moving desks or is that on the client?
    - Are you disconnecting electronics or should IT handle that?

If it lands on you:

- Take pictures of desk setups **before** you move anything
- Label wiring and cables with tape and markers
- Move furniture carefully—don't drag heavy items over floors you're not replacing

When you put things back:

- Get it close to how it was
- Plug in critical stuff if possible (or label clearly if not)

You're not an IT tech, but you also don't want to be the reason someone's system dies for a week.

Work Hours

SOME OFFICES WILL:

- Clear full areas for you
- Let you work after-hours or weekends

Others will want:

- "Half the office done while the other half keeps working"

In split situations:

- Use temporary barriers or sections to keep you out of their daily flow
- Coordinate noise and dust so they're not on Zoom calls over a grinder
- Keep your pathways neat and your gear contained

Residential: You're in Someone's Home

WORKING IN HOMES IS its own world.

You're installing carpet, but the homeowner is:

- Letting strangers into their private space
- Trusting you around their kids and pets
- Watching every move (or feeling nervous in the other room)

First Impressions

SMALL THINGS MATTER:

- Show up when you said you would (or communicate if you're delayed)
- Park where they ask (not blocking neighbors or mailboxes if you can help it)
- Greet them respectfully, make eye contact, listen

Before touching anything:

- Walk the space with them
- Confirm:
    - Areas to be done
    - Furniture plan
    - Pets' situation
    - Bathroom you're allowed to use (if any)

Pets and Kids

PETS:

- Ask if they'll be contained
- Make sure doors and gates that should stay closed do stay closed
- Don't trust "Oh, she never runs out"—that's famous last words

Kids:

- Watch tools—no blades left where a kid can grab them
- Don't leave open adhesive or buckets where a toddler can wander into them
- Keep language clean and behavior calm

Homeowners remember how safe it *felt* having you there.

Respecting Their Stuff

EVEN IF THE HOUSE IS cluttered:

- Don't complain about it to them
- Don't throw personal items onto beds or in piles carelessly
- Use common sense—move things gently, keep like items together

If you find:

- Cash, jewelry, or obviously private items in your way

...don't touch them more than necessary. Ask the homeowner to move them.

Trust is fragile. Don't give them a reason to doubt you.

Cleanup: Leaving It Better Than You Found It

IN OCCUPIED SPACES, **cleanup is half the job**.

No matter how good the carpet is, if you leave the place a mess, they'll remember that, not your seams.

Daily Cleanup

AT THE END OF EACH day, not just at the end of the job:

- Sweep and vacuum your work area
- Remove trash, scraps, and used blades
- Coil cords, store tools, and clear exit paths
- Make sure no tripping hazards are left overnight

Final Cleanup

WHEN THE JOB IS DONE:

- Vacuum the carpets you installed (or at least clean up obvious debris)
- Wipe adhesive, marks, or smudges off nearby surfaces if you caused them
- Take trash with you or use designated dumpsters (not random cans)

Then:

- **Do one last walk-through** with the client if possible
- Ask, "Is there anything you see that we need to address before we go?"

Sometimes they'll say no. Sometimes they'll point out something minor you can fix in five minutes. Either way, you've just shown them:

"I care about how I leave your space, not just how fast I can leave."

Job Note – The Condo Board That Almost Fired Me

I DID A HALLWAY JOB in a condo building occupied, lots of older residents, and a board that treated every little detail like it was a federal case. It was the kind of place where every decision went through a committee, and every complaint came with a photo.

We had a solid plan going in. The work was phased floor by floor, notices went on doors, and we agreed on daytime hours only. On the install side, we were doing fine: prep was good, seams were clean, stretch was tight. The problem wasn't the carpet. It was everything around it.

On that first day, we tracked a little dust into the elevator, left a couple pieces of cardboard near the trash room "just for now," and leaned some tools in the corridor between trips. Nothing wild. Nothing we hadn't seen on a hundred other jobs. But in a building full of people walking those halls all day, it looked sloppy.

By the afternoon, one resident had taken pictures of everything and emailed the board with the subject line: "These guys are a mess." That email made it to the property manager, who pulled me aside and said, "If this is how it's going to go, we're going to stop the job."

The hall we'd finished looked good. The carpet itself was solid. But the way we were living in their common space—dust, scraps, tools parked where people walked—made it hard for anyone to see the quality of the install. We weren't just being judged on our seams; we were being judged on our footprint.

We regrouped fast. One guy on the crew became the dedicated "clean runner." We laid extra protection in the elevator and lobby. At the end of every day, we swept, vacuumed, and walked the whole corridor before we left. Nothing stayed in the common areas unless we were actively using it.

That job taught me that in occupied buildings, a clean, controlled jobsite is part of the installation. The carpet can be perfect, but if people have to step over your mess to see it, all they'll remember is the mess

Checklist – Working in Occupied Spaces

KEEP THIS IN YOUR HEAD (or on your dash) for any job where people are still living or working around you.

Before the Job

- Do I know working hours and noise restrictions?
- Do I know where to park, enter, and stage materials?
- Have I clarified who moves furniture and who handles electronics?
- Have I explained, in simple terms, what the disruption will look like?

Protection & Setup

- Pathways protected where we move tools/rolls over finished floors
- Dust control plan in place for grinding/scraping
- Furniture and sensitive items covered or moved safely
- Clear boundaries set for work areas vs. "their" areas

Behavior

- Crew understands this is someone's home/business, not a warehouse
- Noise kept under control—no blasting radios, no shouting matches
- Language and behavior appropriate around staff, guests, kids, and residents
- Bathrooms and break areas used respectfully

Daily Cleanup

- Work areas swept or vacuumed at end of day
- Trash, scraps, and used blades removed
- Tools and cords coiled and stored, not blocking exits
- Hallways and access paths checked for safety and cleanliness

Final Handoff

- Final vacuum or cleanup on new carpet
- Transitions and edges checked for safety
- Client walked through (if possible) and given a chance to point out concerns
- Any small touch-ups handled before leaving for good

If you can check most of those boxes, you're not just a good installer—you're someone people are happy to invite back.

And in the long game, that's where repeat work, referrals, and better jobs come from.

Chapter 10 – Choosing the Right System for the Job

Most installers are trained to be loyal to a **method**:

- "We stretch everything over pad."
- "We glue everything down."
- "We're a coating crew. If it ain't epoxy, we don't care."

That works—until it doesn't.

You get called into a space that:

- Has high moisture
- Gets hammered by carts and chairs
- Needs quiet but can't have thick cushion
- Or is going to be reconfigured every few years

If you only have one favorite system, you either:

- Force a bad match and hope it survives, or
- Walk away from good work because "it's not what we do"

This chapter is about something bigger:

Choosing the right system for the job, not the right job for your favorite system.

You don't need to be an expert in every product on earth. But you do need to understand:

- What each main system is good at
- Where it struggles
- And how to steer owners toward floors that will do their job

Start with the Space, Not the Material

BEFORE YOU THINK "CARPET vs. LVT vs. coating," ask some questions about the **space** itself.

1. What kind of traffic and abuse will it see?

- Feet only?
- Rolling chairs, carts, wheelchairs?
- Forklifts or pallet jacks?
- Pets, strollers, cleats?

HEAVY POINT LOADS AND rolling loads usually:

- Hate cushion and soft backings
- Love hard surfaces or dense, properly supported systems

Light foot traffic:

- Can handle more softness, more texture, more pattern play

2. What's the moisture story?

- Below grade or on grade slab?
- Known moisture issues?
- Visible efflorescence or past failures?

SOME SYSTEMS:

- Tolerate minor moisture with the right adhesive or underlayment
- Others need a dry, controlled slab or a separate moisture mitigation system

You can't "will" carpet or luxury vinyl to survive a slab that's bleeding water.

3. What's the sound and comfort requirement?

- Is it a quiet office or a call center?
- Hotel corridors over sleeping rooms?
- Apartments or condos with people below?
- Or is it a warehouse where nobody cares how loud it is?

SOUND AND COMFORT OFTEN push toward:

- Carpet (stretch-in, double-stick, or tile)
- Underlayment under LVT
- Resilient rubber in gyms

Bare concrete with epoxy is beautiful and tough—but it's loud and unforgiving.

4. How is the space cleaned and maintained?

WHO'S CLEANING:

- A professional night crew with autoscrubbers?

- A small staff with mops and vacuums?
- A homeowner with a vacuum and a Swiffer?

Some systems:

- Need specific chemicals and procedures
- Require periodic re-coats
- Or react badly to harsh cleaners

If the maintenance reality doesn't match the manufacturer fantasy, the floor will age ugly.

5. How often will the space change?

- Is it a long-term design (like a hotel corridor that might stay the same for 10–15 years)?
- Or a flex office that reorganizes every 3–5 years?
- A retail space with frequent layout updates?

SPACES THAT CHANGE a lot:

- Benefit from systems that are easy to replace in sections (carpet tile, LVT, modular products)
- Don't love monolithic, hard-to-repair systems without a plan

The Big Carpet Systems: Stretch-In, Glue-Down, Double-Stick, Tile

LET'S HIT THE MAIN carpet families and what they're good at.

Stretch-In Over Pad

WHAT IT IS:

Carpet stretched over cushion, secured to tackstrip at the perimeter.

Shines in:

- Residential (living rooms, bedrooms, stairs)
- Some hospitality (hotel rooms, certain corridors if details allow)
- Spaces where comfort and warmth matter

Pros:

- Soft underfoot and quiet
- More forgiving over minor subfloor imperfections
- Easier to replace just the carpet later if pad is still good

Cons:

- Not great for heavy rolling loads (chairs, carts)
- Thicker buildup can cause transition headaches at tile/LVT
- Requires good tension and craftsmanship to last

Use it when:

- Comfort and sound control matter

Direct Glue-Down Carpet

WHAT IT IS:

Carpet glued directly to prepared subfloor (concrete, wood, properly approved underlayment).

Shines in:

- Commercial offices
- Retail spaces
- Healthcare corridors (with right product)
- Areas with lots of rolling traffic

Pros:

- Stable under carts and chairs
- Lower profile, easier transitions to other hard surfaces
- Can be easier to clean with commercial vacuums and extractors
- Less risk of wrinkles if installed correctly on a solid substrate

Cons:

- Subfloor flatness and cleanliness are critical, telegraphing is real
- Harder underfoot and louder than cushion systems
- Removal down the road can be tough if aggressive adhesive is used
- Moisture issues in the slab can wreck the glue bond

Use it when:

- Rolling traffic is a big deal
- The owner wants a "commercial clean" look and feel
- The subfloor is suitable and can be prepped properly

Double-Stick / Cushion-Back Systems

WHAT IT IS:

Pad glued to the subfloor, then carpet glued to the pad; or cushion-backed carpet glued directly.

Shines in:

- Hospitality corridors and public spaces
- Higher-end offices
- Places wanting comfort + performance (but with a budget and spec to match)

Pros:

- More comfort and sound control than straight glue-down
- Better performance for rolling loads compared to stretch-in pad systems (with proper product)
- Premium feel underfoot

Cons:

- More expensive (materials + labor)
- Demanding on prep and adhesive choice
- Can be harder to repair if bond fails or moisture intrudes
- Needs installer who follows the manufacturer system—not guessing

Use it when:

- Owner wants "hotel corridor" feel or high-end performance
- There's a real budget to support it

If you're not set up for double-stick, don't pretend. Either get trained or pass it to someone who is.

Carpet Tile

WHAT IT IS:

Modular carpet squares or planks, usually installed with releasable adhesive or tabs.

Shines in:

- Offices
- Schools
- Retail
- Spaces that reconfigure often or need fast repairs

Pros:

- Easy to replace damaged tiles
- Great for phased work (half room at a time, keep business running)
- Designs can play with direction, patterns, and zones
- Often more forgiving for access floors and raised systems

Cons:

- Requires careful attention to direction and layout
- Bad subfloor prep still shows (edges telegraph over ridges and dips)
- Cheap tiles can look cheap fast
- Seams and direction shift can be obvious if poorly planned

Use it when:

- The client values flexibility, quick fixes, and phased installs
- There's potential for damage (chairs, stains, etc.)
- The aesthetic fits modular patterns or planks

When Carpet Isn't the Answer

CARPET IS GREAT, BUT it's not a Band-Aid for every slab.

There are times you should steer toward:

- **LVT / LVP (Luxury Vinyl Tile/Plank)**
- **Sheet vinyl**
- **Rubber flooring**
- **Epoxy / urethane coatings**
- Or a combo system

LVT / LVP

BEST FOR:

- High-traffic areas needing a "wood" or "stone" look without the fragility
- Rental units, condos, commercial corridors, retail
- Spaces where easy cleaning is a must

Pros:

- Durable wear surfaces
- Great for rolling loads (when installed right)
- Water-resistance (not invincible, but better than carpet in spills)
- Modular replacement possibilities

Watch out for:

- Poor subfloor prep telegraphing every ridge and patch
- Moisture issues under click systems and certain adhesives
- Sound—often needs underlayment in multi-family or above-grade work

Sheet Vinyl

BEST FOR:

- Healthcare rooms
- Labs and clean spaces
- Bathrooms and wet areas
- Some commercial kitchens (specialty products)

Pros:

- Fewer seams when planned right
- Good cleanability and resistance to many chemicals
- Welded seams can give near-monolithic surface

Cons:

- Very unforgiving to bad prep
- Pattern layout and seam planning are critical
- Repairs can be tricky to hide

Use it when:

- Hygiene, cleanability, and seam control matter more than softness
- You or your partner crews know sheet layout and heat-weld work

Rubber (Roll or Tile)

BEST FOR:

- Gyms
- Weight rooms
- Some play areas
- Certain industrial or "back of house" spaces

Pros:

- Impact absorption
- Slip-resistance (depending on texture)
- Tough under rolling and heavy static loads

Cons:

- Smell, especially with some recycled rubber, during and right after install
- Needs proper adhesive and prep or it'll curl, bubble, or shrink
- Thickness and density choices matter a lot

Use when:

- Impact and safety are top priorities
- Client understands cleaning requirements (no mopping with whatever is in the closet)

Epoxy / Urethane / Fluid-Applied Coatings

BEST FOR:

- Garages
- Commercial kitchens (with specialty systems)
- Industrial areas and warehouses
- Showrooms and auto spaces

Pros:

- Monolithic, seamless look (when done right)
- Chemical and abrasion resistance with the right system
- Easy to clean with proper procedures

Cons:

- Slippery if the system or texture is wrong for the environment
- Very demanding on concrete prep and moisture control
- Mild odors/chemicals during install (ventilation and PPE required)
- Not forgiving of slab movement or poor prep

Use when:

- Client wants "concrete, but better"
- They get that it's firm, loud, and not soft like carpet
- Budget allows real prep (grinding, moisture work if needed)

Matching Systems to Clients, Not Just Spaces

YOU'RE NOT JUST MATCHING material to concrete. You're matching systems to humans.

Ask:

- Who is this client really?
- What do they care about?
- Where will they **for sure** cut corners if you let them?

Examples:

The aesthetics-first client (designer / high-end homeowner)

- Cares deeply about how it looks
- May not be thinking about maintenance or life-cycle
- You need to explain how their beautiful choice will live in the real world

The budget-first client (small business / tight HOA)

- Wants the number low
- Might ask you to skip prep, cushion, or quality materials
- You need to explain what cutting those things will cost them later

The operations-first client (facility manager / hotel GM)

- Cares about uptime, safety, cleaning
- They'll listen if you talk in terms of:
    - Fewer rooms out of service
    - Easier cleaning
    - Fewer complaints

Your advice has more power when it speaks their language.

Steering Away from Bad Choices (Without Starting a Fight)

SOMETIMES THE SPEC or client request is obviously wrong for the situation.

Examples:

- Thick, plush carpet in a rolling office environment
- High-shine epoxy in a wet, greasy commercial kitchen with no texture
- Cheap click LVT on a slab with known moisture issues

Instead of, "This is stupid," try:

"We *can* install that, but here's what I've seen happen in spaces like this..."

Then spell out consequences:

- "Those chairs will chew up this style of carpet fast."
- "This shiny finish will be slippery when it's even a little wet."
- "With your moisture readings, this click product will likely cup or gap."

Offer alternatives:

- "If you want soft but need rolling strength, we could do cushion-back carpet tile."
- "If you want this look but safer, we can add texture or flakes for grip."
- "If we stay with LVT, I recommend this glue-down system with a moisture-tolerant adhesive and mitigation."

You're not trying to win an argument. You're trying to:

Give them clear, honest choices with real consequences.

Job Note – The Corridor That Needed Tile, Not Carpet

I WAS CALLED TO LOOK at a long corridor in a medical office building. The manager walked me down the hallway and kicked at the old, matted carpet like it had personally offended him. He said what a lot of people say in that situation: "We're tired of this stuff. We want something softer and warmer than tile, but we need to keep it cheap."

The space looked simple at first glance—just a straight run of corridor. But as we walked, the real story showed up: constant rolling traffic from carts and wheelchairs, staff hustling back and forth all day, regular spills, and that sharp disinfectant smell that tells you the cleaners are not gentle.

What they wanted was basic stretch-in carpet over pad. It was the cheapest option on paper, and they'd had "regular carpet" before, so in their minds this was just doing the same thing again, hopefully nicer.

Everything in me said that was the wrong system for that job.

Soft pad and stretch-in in a medical corridor like that is a recipe for waves, crushed traffic lanes, and a short, ugly lifespan. The cleaning chemicals alone were going to beat on the backing and pad. The rolling loads would find every bit of slack and turn it into a wrinkle. They'd be calling someone like me back in a year or two, mad at the carpet and confused about why it failed again.

I didn't tell them they were stupid. I just laid out what I'd seen on other jobs like theirs. I told them if they really wanted the best balance of comfort, durability, and maintenance, we should be talking about carpet tile or a good commercial glue-down system, not cheap pad and stretch-in.

They didn't like the price difference at first. We went back and forth. Finally I framed it the way that mattered to them: they could save money right now and pay for it again in headaches and replacements or spend more up front on a system that would handle their traffic and cleaning without falling apart.

In the end, they went with carpet tile—more expensive on day one, cheaper over the life of the corridor.

Three years later I came back to look at another area in the same building. I walked that hallway again. The one tile in the hallway had been swapped once because of a spill, but the field was still flat, tight, and respectable. Staff weren't tripping over wrinkles or complaining about it. The manager told me, "I'm glad we listened, it still looks beautiful."

That office job reminded me why you can't just install whatever people ask for and keep your mouth shut. Part of the job is steering owners away from repeating their own bad choices. You won't win every one of those conversations, but when you do, the floor—and your name—hold up a lot better over time.

Checklist – Choosing the Right System

USE THIS BEFORE YOU jump to, "We'll just do what we always do."

Space & Use

- Do I understand traffic type and intensity (feet vs carts vs forklifts)?
- Do I know how wet, messy, or clean this space tends to be?
- Do I know whether sound and comfort are priorities here?

Substrate & Environment

- Is the slab or subfloor in decent shape—or does it need serious work?
- Are there known or suspected moisture issues?
- Are there level changes or transition challenges to nearby surfaces?

Operations & Maintenance

- Who cleans this floor and how (pro crew vs homeowner vs staff)?
- Are they set up to maintain the system I'm proposing (chemicals, equipment)?
- How often are they likely to replace or reconfigure the space?

System Match

For the options on the table, ask:

- Does this system fit traffic and abuse?
- Does it match their sound/comfort expectations?
- Is the subfloor and moisture situation acceptable for this system?
- Can they realistically maintain it?

Client Fit

- Have I explained trade-offs in their language (comfort, cost, uptime, aesthetics)?
- Have I pointed out obvious risks if they choose the "wrong" system anyway?
- Have I offered at least one alternative that fits better?

If you can work through those questions honestly, you're no longer just "the carpet guy."

You're:

- A systems thinker
- A guide
- Someone owners can trust when they're about to spend real money on something they'll walk on every day

That's how you get called back not just to **install**, but to **advise**.

In the next chapter, we'll shift from materials and methods to **money and margins**—how to price your work so you're not just staying busy but staying in business.

Chapter 11 – Pricing, Estimating, and Your Margins

There are two kinds of "busy" in this trade:

Busy and profitable / Busy and broke

From the outside, they look the same:

Van moving, phone ringing and jobs on the schedule.

From the inside, they feel very different.

Most installers don't stay in the game because of their seams.

They stay—or burn out—because of their **pricing**.

This chapter is about:

- Knowing your numbers
- Building estimates that make sense
- Saying no to work that will eat you alive
- And protecting your margins so you're not just surviving, but building something

You're not just installing floors.

You're running a business every time you write a number down.

Step 1 – Know What an Hour of Your Time Really Costs

BEFORE YOU TALK ABOUT rates and bids, you need to know **your own cost**.

Most installers do this:

"So-and-so charges X per yard. I'll be a little cheaper."

That's not a strategy. That's a slow-motion disaster.

The Real Cost of an Hour

WHEN YOU (OR YOUR CREW) are on a job, the cost isn't just "what you'd like to make."

You're paying for:

- Your own wage (what you want to take home)
- Helpers' wages
- Payroll taxes and fees
- Fuel
- Vehicle payment/maintenance
- Insurance (liability, auto, maybe workers comp)
- Tools and repairs
- Shop rent / storage
- Phone, software, bookkeeping, website
- Random "stuff happens"

You don't need a fancy spreadsheet. But you *do* need a realistic number.

Example idea:

- Figure out what you'd like each **working** year to pay:
    - Your take-home target (say, $70k)
    - Overhead (say, another $30k)
    = $100k that your work has to generate just to exist
- Now divide that by **billable** hours:
    - You might work 2,000 hours in a year
    - But not all of that is billable (drive time, estimates, downtime)
    - Maybe you realistically bill 1,400–1,600 hours

$100,000 ÷ 1,500 hours ≈ $66/hour just to hit that target.

That's **before** profit.

You don't have to use those exact numbers. The point is:

You need to know a real "hourly cost" baseline,

or every bid is just guessing.

Once you know that number, you can:

- Translate it to per-day, per-yard, or per-foot pricing
- See fast when a job will work—or won't

Step 2 – Break the Job into Pieces, Not Just Square Footage

THE WORST PRICING HABIT:

"Room is X by Y, so that's Z yards, times my rate, done."

That might get you close on some simple jobs.

But good estimating means **breaking the job down**.

Ask: What am I doing here?

- **Demo & disposal**
    - Take-up, haul-off, dump fees, time up/down stairs or elevators
- **Subfloor prep**
    - Scraping, grinding, patching, self-leveling (if in your scope)
- **Installation**
    - Field work, stairs, details, transitions
- **Occupied space factors**
    - Moving furniture
    - Night/weekend work
    - Extra protection and cleanup
- **Materials**
    - Carpet or product (if you supply)
    - Pad, tackstrip, adhesive, tape, patch, metals, transitions, fasteners
- **Travel and logistics**
    - Drive time, parking, loading/unloading challenges

You don't have to write a novel every time. But when you look at a job, your brain should go:

"This isn't just 120 yards — this is 120 yards + 40 yards of demo + patching + stair work + two trips." Then your number starts making sense.

Step 3 – Choosing How You Price: Yard, Foot, Day, or Lump Sum

THERE'S NO ONE "RIGHT" way to price everything. Different jobs run better on different structures.

Per Yard / Per Square Foot

GOOD FOR:

- Straightforward carpet installs
- Repetitive work (hotel rooms, tract homes, similar offices)

Pros:

- Easy for clients to understand
- Fast to calculate on simpler jobs

Cons:

- Can bury complexity (stairs, details, occupied work)
- Tempts you to ignore weirdness because "it's just X per yard"

Use per-yard/foot when:

- Scope is clear
- Space is simple
- You add **separate line items** or adders for tricky areas

Per Day / Per Crew Day

GOOD FOR:

- Unknowns ("This place is a mess; we don't know exactly what we'll find.")

- Prep-heavy jobs
- Projects where you're more of a partner than a commodity

Pros:

- Protects you when the work expands in ways you warned them about
- Keeps the focus on *phases*: "Phase 1: Demo and prep – X days"

Cons:

- Some clients hate "day rates"—they want fixed numbers
- You have to be honest and efficient, or they'll feel taken advantage of

Use day rates when:

- You've explained clearly what can change
- The client understands you're not milking time, you're **defending the schedule and quality**

Lump Sum

GOOD FOR:

- Well-defined scopes
- Commercial bids
- When the client wants "one number"

Pros:

- Simple for the client
- Lets you make more if you work efficiently

Cons:

- If you guessed wrong, you eat it
- Scope creep will destroy you if you don't manage changes

Use lump sum when:

- You've walked the job
- You've thought through every major step
- You're ready to say, "That's not in the original scope" if they add things later

Step 4 – Adders: Don't Do Extra Work for Free

SOME THINGS SHOULD almost **always** bump your price.

If you don't account for them, they quietly erase your margins.

Common adders:

- **Stairs**
    - Per step rate or per flight rate
    - More for cap-and-band / Hollywood than simple waterfall

- **Patterns**
  - Extra for pattern match, repeats, layout time
- **Double-stick / cushion-backed**
  - Extra materials, adhesives, time, and risk
- **Occupied spaces**
  - Furniture moving
  - Off-hours or night work
  - Extra protection and cleanup
- **Heavy demo / glue scrape**
  - Old double-glued pad
  - Thinset ridges
  - Layered flooring history
- **Travel**
  - Jobs an hour away or more, especially for small scopes

You don't have to charge gold-plated numbers. But you **do** have to:

- Recognize what takes more time and risk
- Capture that in your price

Otherwise, you're effectively doing those things for free.

Step 5 – Guardrails: What's Not Included (Unless They Pay)

HALF OF PROTECTING your margins is about being clear about what you **are not** including.

Examples:

- Major leveling or concrete repair beyond normal skim patch
- Moisture mitigation systems
- Plumbing, electrical, or IT work
- Moving extremely heavy items (pianos, safes, full wall units)
- Repairing damage caused by other trades or existing conditions

You can say it simply in your proposals:

- "Price includes standard prep (scraping, minor patching). Extensive leveling or moisture mitigation, if required, will be an additional charge."
- "Furniture moving includes standard items only. Pianos, safes, large aquariums, and built-ins are excluded."

You're not trying to nickel-and-dime them. You're:

- Drawing a line between a normal job
- And a problem someone wants you to fix for free

Step 6 – The "Cheaper Guy" and How Not to Race to the Bottom

THERE IS ALWAYS A CHEAPER guy.

If your whole pricing strategy is:

"I'll undercut whoever they mention by a little bit,"

...you're building a business on knees and hope.

When a client says:

- "We have another bid that's lower. Can you match it?"

Instead of panicking, try:

"Sometimes I'm not the cheapest.

My price includes proper prep, correct adhesives, and making sure your floor lasts.

If their number doesn't include that, they're not cheaper—they're just leaving things out."

You can ask:

- "Does their price include...?"
    - Demo and disposal?
    - Subfloor prep?
    - Pattern match?
    - Night work?
    - Warranty on labor?

If it does, okay—they might just be underpricing themselves. You can decide:

- Do I want this job **at that number**?
- Or am I willing to walk?

If it doesn't, then:

- You're not losing to a cheaper *equivalent*
- You're choosing not to match a **worse** offer

You don't have to win every job. You must win the **right** ones.

Step 7 – Scope Creep and Change Orders: How to Say "Yes, But..."

ON ALMOST EVERY JOB, someone will say:

- "While you're here, could you also...?"

Small things add up. If you say yes to everything for free, you're giving away days of work.

A simple way to handle it:

1. **Acknowledge the request**
    a. **"Yes, we can take a look at that."**
2. **Separate it from the original scope**
    a. **"That's outside what we quoted for this job."**
3. **Give them options**
    a. **"If you'd like, I can add that for X dollars."**
    b. **"Or we can come back another day and give you a separate quote."**
4. **Get agreement before doing the extra**
    a. **Even a text or email confirmation is better than nothing:**
        i. "Just confirming we're adding the back room for $___."

You're not being difficult. You're being clear.

If you occasionally do a **tiny** thing for free as goodwill, that's okay. Just make sure:

- It's your choice, not their assumption
- You don't train them to expect unlimited extras

Step 8 – Deposits, Draws, and Getting Paid

WE'RE NOT DOING LEGAL advice here, but we can talk common sense.

You need cash flow.

Habits that help:

- **Deposits on material jobs**
  - Enough to cover materials and some labor upfront
  - Especially for custom orders
- **Progress draws on bigger projects**
  - Don't wait 60 days to get paid for a month's worth of work
  - Set milestones: demo/prep, mid-install, final
- **Clear payment terms**
  - "Due on completion" means at the end of the job, not 45 days later
  - Decide how you handle late payments (fees? stop work?)
- **Paper trail**
  - Written estimates, change orders, and invoices
  - Keep texts/emails where you agree on scope and price

You don't have to be aggressive. But you also don't have to be a bank.

If a client is always late, always arguing about pennies, or always "forgetting" changes they asked for, ask yourself:

"Is this someone I want as a repeat customer?"

Job Note – The Job That Paid Me About Two Bucks an Hour

EARLY ON, I LANDED on what I thought was a dream project.

Big square footage. Nice space. Good-looking plans. The kind of job you imagine putting in your photo book and bragging about for a while. I wanted it bad enough that I made the classic mistake: I chased it with my emotions instead of my numbers.

I sharpened my pencil on the install price to beat everyone else. Then I sweetened the deal by throwing in demo "to make the bid attractive." I included some floor prep without really defining what "basic prep" meant. I didn't add anything for working around other trades or dealing with the building's weird access. On paper, it still looked okay: plenty of yards at a decent rate. Easy money, right?

The reality showed up the first week.

The old floor came up hard. There was double-glued pad in some areas, mystery mastic in others, and layers of history nobody mentioned. Scraping and demo took twice as long as I'd guessed. The subfloor was wavy, patched, and full of surprises, so we spent extra time just getting it to a place where I could install without hating myself.

Other trades were in our way, so a lot of our day turned into waiting, shuffling, and redoing protection. The "simple layout" wasn't simple once real walls, doors, and furniture showed up. On top of that, the client kept saying, "While you're here, can you just take care of this area too?" and I kept saying yes because I wanted to keep them happy.

By the time we were done, the job looked good. The client was thrilled. They talked about more buildings down the road. On the install side, it was a success.

Then I sat down and did the math.

When I added all the hours for me and my helper, all the extra trips, the extra prep, the throne-in demo, and the free "while you're here" extras, the rate I had effectively paid myself worked out to something like two dollars an hour. It was a punch in the gut. I'd worked my tail off and fallen behind on other opportunities... for less than what a teenager makes bagging groceries.

The worst part? The client wanted the next job "at about the same price as this one."

That's when I realized the trap I'd built for myself. I'd trained them to expect top-tier work at bottom-tier numbers. If I kept playing that game, every "dream project" was going to be another long week of donating my body to somebody else's margin.

On the next conversation, I told them the truth. I said, "I'm glad this turned out well for you, but at the price we did this first one, I lost money. If we're going to do more buildings, the price has to match the actual work." Some people would have walked at that point. To their credit, they didn't. We sat back down and rebuilt the numbers with real demo, real prep, and realistic labor in the mix.

That job taught me never to discount my price based on how badly I want the work. If the numbers don't make sense—after you count the hours, the headaches, and the risk—it's not a good job for you, no matter how pretty the drawings look or how nice the building is.

Checklist – A Quick Margin Check Before You Send a Bid

BEFORE YOU HIT "SEND" or write that number on the paper, run through this mentally:

1. Have I...

- Counted *all* the work? (demo, prep, install, details, travel)
- Added something for stairs, patterns, double-stick, or tricky areas?
- Included the headache factor of occupied spaces, nights, or tight timelines?

2. Do I...

- Know roughly how many **hours/days** this will take?
- Know my **real hourly/day cost**, not just my "wish" number?
- See a margin on top of cost, not just "break-even if nothing goes wrong"?

3. Have I...

- Told them what is **not** included (major leveling, mitigation, heavy moving, etc.)?
- Planned how extras or scope changes will be handled (change orders, add-ons)?
- Set clear payment expectations (deposit, draws, due-on-completion)?

4. Am I...

- Okay with this client at this price if something goes wrong?
- Prepared to walk away if they demand you cut below your survival level?

If you can check most of those boxes and the number still feels right, you're in good shape.

If you find yourself thinking, "Well, ...that's not a bid—that's a prayer."

You don't have to be the most expensive.

You don't have to be the cheapest.

Your goal is:

To price your work so you can do it **right**,

stand behind it,

and still have money and energy left when the job is done.

That's how you build:

- A reputation
- A real business
- And a life outside the van

Part IV – Career & Mindset

Chapter 12 – Reputation and Playing the Long Game

You can be the best installer in your city and still be broke and exhausted if:

- Nobody knows you
- The wrong people know you
- Or the only people who know you are the ones who want everything cheap, fast, and yesterday

Skill gets you **through** the door.

Reputation decides **which** doors open and **how** you're treated when you walk in.

This chapter is about:

- The kind of name you're building
- The kind of clients you attract
- And how to turn "one job" into a pipeline instead of a one-and-done grind

You're not just building floors.

You're building a story people tell about you when you're not in the room.

What Do You Want to Be Known For?

IF YOU DON'T DECIDE, the world will decide for you—and it might not pick what you'd like.

Common default identities:

- "The cheap carpet guy"
- "The guy who will always squeeze us in at the last minute"
- "The magician who fixes everyone else's screwups"
- "The quiet pro who does it right and doesn't cause drama"

You can't control every story, but you *can* aim at one.

Ask yourself:

- Do I want to be known first for **price**... or **result**?
- For being **fast and sloppy**, or **reliable and solid**?
- For taking **any job**, or taking the ones that fit what I do best?

You don't have to write a mission statement. Just settle on something like:

"We're the crew you call when you want it done **right**, not just cheap."

Or:

"We're the people-friendly flooring crew that respects your space and your schedule."

Once you know that, you can start acting and talking like it—on purpose.

Your Reputation Is Built in Tiny Moments

MOST PEOPLE WON'T CRAWL around inspecting your seams.

They will remember:

- Whether you showed up when you said you would
- Whether you left the place clean
- Whether you owned your mistakes
- Whether you made them feel dumb... or respected

Some of the tiny things that add up:

- Calling if you're running late, instead of just showing up whenever
- Protecting their floors and furniture without being asked
- Saying "We'll take care of that" when you obviously caused a small issue
- Not badmouthing other installers in front of the client (even if they deserve it)

None of that takes special skill. It takes **choices.**

Those choices turn into:

- Reviews
- Referrals
- Or silent blacklists you never hear about

Pick Your Clients... or They'll Pick You

EARLY ON, YOU TAKE whatever comes. But at some point, you must start asking:

"What kind of clients do I actually want more of?"

Red-Flag Clients

THESE ARE THE ONES that burn you out.

Common signs:

- They start the conversation with:
    - "We had a horrible experience with the last five installers..."
    - "I just need someone to beat this price."
    - "This job will lead to lots more work" (but they're fighting you over your first quote already)
- They ignore what you say about:
    - Moisture
    - Prep
    - Realistic timelines
- They try to negotiate every line item like it's a flea market

Red flags don't mean they're evil. It just means:

- They're not ready for the way you work
- Or they want a level of discount and control that kills your margin

You can still help them—by being **clear**:

- "If you're just looking for the cheapest number, I'm probably not your guy.
  If you want it done right, here's how I do it and what it costs."

Some will walk. That's okay. You just avoided months of stress for low pay.

Green-Flag Clients

THESE ARE THE ONES you quietly wish you had 50 of.

Signs:

- They ask good questions and listen to the answers
- They care about the floor lasting, not just "Can you be here Tuesday?"
- They answer emails and approve decisions in a reasonable time
- They pay on time without turning every invoice into a debate

Often:

- Facility managers
- Good GCs who value their subs
- Homeowners who respect trades
- Small business owners who "get it" because they've been burned before

When you find them, do more than just say thanks. Learn:

- What they care about
- How they like to communicate
- How you can make their life easier next time

These are the people who will quietly fill your calendar if you treat them right.

How to Turn One Job into Many

EVERY PROJECT HAS THREE phases:

1. Getting it
2. Doing it
3. What happens **after**

Most installers focus on phase 2.

The long game is mostly in phase 3.

Ask for the Right Kind of Review

YOU DON'T HAVE TO BEG. At the end of a job that went well, you can say:

- "If you're happy with how this turned out, a short review or reference would really help me out."

Then make it **easy**:

- Send them a link (Google Business, website, wherever you want reviews)
- Or say, "If someone asks you who did your floors, feel free to pass my name along."

You're not chasing stars for ego. You're:

- Making it easier for good clients to find you
- Putting proof behind your claims

Stay in Touch Without Being Annoying

YOU DON'T NEED A FANCY marketing plan. Simple ideas:

- When you drive past a big job you did a year ago, shoot a quick text:
    - "Hey, just drove by and thought of this job—hope the floors are still treating you well. Holler if you need anything."
- Once or twice a year, send key clients a short check-in:
    - "Any flooring issues or future projects I can help you think through?"

You're not selling hard. You're:

- Staying top-of-mind
- Being the person they think of *before* they call some random number online

Ask for Referrals Like a Pro

AFTER A GOOD JOB, ESPECIALLY with green-flag clients:

- "If you know any other managers or owners who struggle with flooring issues, I'm always glad to share what I know—even if they don't hire me."

That frames you as:

- A resource
- Not just a guy chasing the next deposit

People like referring someone who makes them look smart.

Be that someone.

Online Presence Without Losing Your Mind

YOU DON'T HAVE TO BECOME a full-time influencer. But basic online presence helps:

- People check you out before they call
- Good clients use Google like everyone else

Bare minimum:

- A simple website or landing page with:
  - What you do
  - Areas you serve
  - A few good photos
  - A way to contact you
- A Google Business profile:
  - So reviews can live somewhere
  - So you show up on the map when someone searches "carpet installer near me"

Optional but powerful:

- Before/after photos on a simple gallery or social account
- Short writeups:
  - "Hotel corridor refresh – solved seam issues, reduced noise, night work only"
  - "Garage coating – solved hot tire pickup, easy clean"

You're not trying to impress other installers (though some will peek).

You're trying to tell potential clients:

"This is the kind of work I do.

This is how I think.

This is what your place could feel like."

Partnering with the Right People

YOU DON'T HAVE TO FIND every job yourself.

Some of your best future work will come from:

- General contractors
- Designers
- Property managers
- Material reps
- Other installers who don't do what you do

The key is picking **which** relationships to lean into.

Good GC / PM Relationships

A GOOD GC OR PROPERTY manager:

- Tells you the truth, even when it hurts
- Backs you when you say, "Subfloor is not ready yet"
- Doesn't always pick the cheapest number

You build those relationships by:

- Hitting your dates (or communicating early if something changes)
- Not making everything an emergency
- Helping them look good in front of owners
- Owning your part when something goes wrong

You don't have to be perfect. You must be **reliable**.

Designers and Reps

DESIGNERS:

- Can send you into higher-end spaces
- Want things to look right and match the concept

Reps:

- Know who is buying what
- Need installers who won't make their product look bad

If you become the installer who:

- Reads the spec
- Calls before installing something risky
- Sends photos of finished work

...you'll get more calls when they need someone who won't wreck their job.

Saying "No" (So You Can Say "Yes" to Better Work)

A BIG PART OF THE LONG game is learning to decline the wrong work without burning bridges.

Some jobs you say no to because:

- The price can't get where it needs to be
- The client is a walking red flag
- The conditions are bad and they refuse to fix them
- The schedule is impossible without cheating

The way you say no matters. Example:

"I appreciate the opportunity, but at that price I can't do the job correctly and stand behind it. I'd rather pass than cut corners on you."

Or:

"With the moisture issues and the timeline, I'm not the right fit for this one.

If you ever have a project where we can address those things properly, I'd be glad to talk."

You're:

- Protecting your name
- Leaving the door open for future, better-fit jobs
- Choosing not to become "the guy who does anything at any price"

Every time you say no to a job that would wreck you, you're saying yes to:

- Your sanity
- Your best clients
- The work you want to be known for

Job Note – The Corridor That Sold Me for Five Years

I ONCE DID A CORRIDOR job for a small, well-run hotel. Nothing flashy. No giant corporate budget, no designer hovering over every swatch. Just a clean property with an owner who cared and a GM who wanted the place to feel better than its price point.

The scope was straightforward on paper: tear out the tired old carpet, fix what needed fixing underneath, and install new patterned carpet in the main hallway and a couple of side runs. The catch was that the building was fully occupied. Guests were coming and going. The front desk was nervous about the noise. Housekeeping had their own schedule. It was one of those jobs where the install is only half the work; the other half is how you live in their space while you do it.

We treated it like our own showcase even though nobody called it that. We took the prep seriously instead of just scraping the worst and hoping the pad would hide the rest. We spent extra time on layout so the pattern didn't fight the walls or the light. We stretched it properly with power stretchers, instead of letting the "We've got guests upstairs" pressure talk us into shortcuts. And every night before we left, we swept, vacuumed, and walked the corridors like we were guests ourselves.

There were a few minor issues, like there always are. A door that rubbed. A transition that needed a little more attention. We handled them quickly without making a scene or looking for someone to blame. By the end, the hallways felt calm underfoot, looked clean under the lights, and didn't give you anything to trip over with your eyes or your feet.

As we were wrapping up, the GM said something that stuck with me: "We've had a lot of different trades here. You guys are different." At the time, I took it as a nice compliment and moved on to the next job.

Over the next few years, that corridor kept working for me.

The GM moved to another property and called me there. The regional manager for the ownership group referred me to their sister hotel across town. One of the owners had a small apartment complex and asked me to look at the halls in that building. None of those conversations started with, "What's your price per yard?" They started with, "We know how you treated us last time. We just need to know what it'll take to do it right here."

That one hotel corridor turned into a small pipeline of steady, decent-paying work without a single ad or coupon. All because we treated a "normal" job like it mattered, respected their space while we were in it, and left behind something we weren't afraid to have our name connected to.

That's when it really sank in that a good job isn't just today's paycheck. It's a seed. You can toss it in the ground and forget it, or you can plant it on purpose and let it grow into something that feeds you for years

Checklist – Building a Reputation That Works for You

BEFORE YOU THINK "I just need more jobs," ask:

What am I known for right now?

- Do clients say I show up when I say I will?
- Do they feel I care about their space (not just my task)?
- Do I own and fix my mistakes instead of making excuses?

Am I choosing my clients, or taking whatever hits my phone?

- Do I recognize red-flag clients and at least adjust my price/scope accordingly?
- Do I know my favorite clients by name and what they care about?
- Do I have a way to gently say no when something is obviously a bad fit?

Am I turning one job into many?

- Do I ask for reviews or referrals after good jobs?

- Do I stay in touch with my best clients a couple of times a year?
- Do I thank people when they send work my way?

Does my online presence match who I am on-site?

- Do I have at least a **basic website** or profile that shows what I actually do?
- Do I have a few **real photos** of jobs I'm proud of?
- Are there a few **honest reviews** or testimonials anywhere people can see?

If you can honestly check most of those, then you're building your name and a good name is worth more than any single project.

It's the thing that keeps better clients calling better jobs showing up and better money behind the work you already know how to do

That's how you go from "installer" to **master of your craft and your business**—one job, one conversation, one corridor at a time.

Chapter 13 – Staying in the Game

Most installers think about tomorrow like this:

- "What's on the board?"
- "Where do we start?"
- "Do I have enough glue, pad, and blades?"

That's important.

But there's a bigger set of questions:

- What kind of shape will your body be in 5–10 years from now?
- Will you still be crawling and stretching... or leading and directing?
- Will you have a crew you trust—or just a list of phone numbers?
- When people in your town say your name, what will *that* mean?

This chapter is about the long game behind the long game:

Not just how to install floors well,

but how to build a life, a crew, and a legacy that makes all the hard work worth it.

Your Body Is Your First Tool

YOU CAN REPLACE A VAN.

You can buy more knives, irons, grinders, and stretchers.

You only get one set of:

- Knees
- Back
- Shoulders

- Lungs
- Hearing

If those go, *everything* gets harder.

Knees and Back: "I'll Be Fine" Is a Lie

EVERY INSTALLER HAS said this:

"I'm fine. I'll rest when this job is done."

Then there's always another job.

Habits that quietly destroy you:

- Crawling on bare concrete all day
- Lifting rolls alone because "it's faster"
- Bending from the waist instead of the hips
- Using your back instead of leverage and tools

Habits that save you:

- **Decent knee pads** you wear
- Using dollies, carts, and helpers for big rolls
- Bending your knees when lifting, not just hunching
- Taking 30 seconds to stretch your back and hamstrings a couple times a day

You don't need a yoga routine on the job. You just need:

- Enough movement that your body doesn't lock up
- Enough respect for future-you that you don't treat your joints like they're disposable

Hands, Wrists, and Repetitive Stress

EVERY DAY YOU:

- Grip knives and trowels
- Squeeze staplers
- Push, pull, and twist

Little pains become big problems if you ignore them.

Helpful habits:

- Take a couple minutes before each job to stretch
- Switch hands when possible (even if the "wrong" hand feels dumb at first)
- Use sharp blades so you're not forcing cuts
- Take micro-breaks when you feel tingling or numbness
- Keep your wrists straight when lifting and pulling, not bent at weird angles

If your hands go, the trade becomes a fight instead of a craft.

Lungs and Ears

DUST, ADHESIVES, GRINDING, vacuums, compressors—your lungs and ears take a beating.

Basic protection:

- Wear a dust mask or respirator when grinding, scraping, or working in heavy dust
- Use hearing protection when you know it's loud (grinders, saws, hammer drills)
- Ventilate when using strong-smelling adhesives and coatings

You might feel invincible at 25.

At 45, you'll know whether that was true.

Burnout: When Your Head Is More Tired Than Your Body

BURNOUT SHOWS UP WHEN:

- You're irritable all the time
- You dread the phone ringing
- Every job feels like a fight, even the easy ones
- You start hating the trade you once loved

Burnout doesn't mean you're weak. It means:

You've been giving more than you've been getting back for too long.

Don't Live in Constant Emergency Mode

IF EVERY JOB IS:

- Underpriced
- Rushed
- Full of "just make it work"

...your nervous system never gets a break.

Fixing that comes from:

- **Better estimating** (previous chapter)
- **Better clients** (the one before that)
- Saying "no" to jobs that smell like trouble from the first phone call

You don't need a spa retreat.

You need fewer chaos jobs and more of the ones that fit you.

Boundaries: When Work Ends for the Day

IT'S TEMPTING TO LET work bleed into everything:

- Answering texts at midnight
- Reworking numbers instead of sleeping
- Always being "on call"

Pick some simple rules:

- "No business calls after X time unless it's an active emergency on a job."
- "One day a week with no jobsite work, even if there's paperwork."
- "Family events and key personal things don't always lose to 'one more job.'"

Building a Pro Crew Instead of Just "Having Helpers"

THERE'S A BIG DIFFERENCE between:

- Random helpers you pick up for the day
- A crew that knows your system, your standards, and your way

The first is survival mode.

The second is how you get your life back.

Teach What You Actually Want

"DO IT RIGHT" IS NOT a system your crew needs:

- Concrete standards:
    - "This is what a clean seam looks like."
    - "This is how we protect a client's home."
    - "This is how we leave a job each day."
- Simple processes:
    - Start-of-day setup checklist
    - End-of-day cleanup checklist
    - "Call me before you do X" rules for big decisions

When someone new joins:

- Explain *why* you're doing what you're doing:
    - "We protect this floor first."
    - "We don't leave staples sticking up."
    - "We check for squeaks before pad goes down."

You're not just buying labor hours. You're training **future yous**.

Give People a Way to Grow

IF YOUR CREW SEES:

- No path beyond "grunt with a knife,"
- No raise unless they complain,
- No trust no matter how long they stay,

...they'll either leave or stop caring. You don't need corporate charts. You can say:

- "If you learn layout and start running small jobs, your rate goes up."
- "If you can take a project from start to finish with clean results and no babysitting, you're a lead."
- "If you bring in jobs or relationships, you share in the gain."

It doesn't have to be complicated. It just must be **real**.

Deal with Bad Fits Fast

SOME PEOPLE:

- Won't respect clients
- Won't use tools properly
- Won't show up right, no matter what you pay or say

You can:

- Coach once
- Coach twice

After that, if they're still a problem:

You're keeping them for you, not for the business.

Bad crew members:

- Damage your reputation
- Wear you out emotionally
- Slow down people who *do* care

Letting someone go is hard.

Dragging them along for years is harder.

Systems: So the Job Doesn't Live Only Inside Your Head

Checklists You've seen a bunch in this book.

TURN THEM INTO:

- Printed sheets in your van
- Notes on your phone
- Whiteboard in the shop

Examples:

- "Subfloor prep basics" checklist
- "Occupied job end-of-day" checklist
- "Stairs done right" checklist

Hand them to your crew.

At first, they'll pretend they don't need them.

Then one day they'll quietly pull them out without you asking.

Standard Ways You Do Things

WHERE YOU CAN, STANDARDIZE:

- How you bid certain job types
- How you set up tools in the morning
- How you protect floors and move furniture

The point isn't to turn everyone into robots. It's to:

Free up your brain for the *special* problems

by making the normal stuff automatic.

Planning for "Later" (Even If You're Not Ready to Retire)

ONE DAY, YOU'LL:

- Install less
- Manage more
- Or decide you're done completely

If you've built nothing but calluses, that transition is rough.

You can start now by:

Shifting from "Installer Only" to "Installer + Teacher"

WHEN YOU SHOW A NEW guy how to:

- Row cut
- Run a stretcher
- Read a spec

...you're practicing something bigger:

- Passing on a craft
- Building leadership muscles
- Making yourself slightly less essential to every move

You never stop installing in your head.

But your body doesn't always have to be the first one on its knees.

Building Assets That Aren't Just Your Hands

ASSETS CAN BE:

- A name and reputation that lets you price right
- A small team that can run jobs without you on-site all day
- Written systems and quotes that someone else can follow
- Even a book or guide with your methods stamped on it

You might one day:

- Scale up with more crews
- Stay small but more selective
- Shift into consulting, training, or specialty work

Whatever direction you go, assets = options.

Legacy: What You Leave Behind Besides Carpet

MOST FLOORS YOU INSTALL will eventually be:

- Replaced
- Covered
- Demolished

Your *legacy* isn't the physical carpet.

It's:

- The standards you trained into your crew
- The way clients talk about your name
- The younger installers who think, "I learned the right way from that guy"
- The family that benefitted from you doing this trade on purpose, not just by accident

You don't have to be famous.

You just have to be **consistent**.

One day, someone will ask:

"Who taught you how to do it like that?"

If your name is in that answer, you've done more than just work—you've built something.

Job Note – The Day I Realized I Wanted Out of the Van, Not Out of the Trade

THERE WAS A MORNING I still remember clear as day.

I climbed into the van like I'd done a thousand times before, but everything felt heavier than it should have. My knees were stiff. My back was tight. My fingers ached before I'd even picked up a knife. The schedule for the week was packed, the phone was already buzzing, and all I could think was, *"I'm too tired for it to be this early."*

Nothing was wrong with the job itself. Good client. Fair numbers. Straightforward work. The problem was that I had built a life where **every** job needed me on my knees from start to finish. If I wasn't there stretching, cutting, gluing, fixing, the whole thing felt like it would fall apart. That's a flattering story when you're young. It's a trap when you're worn down.

On that project I had a younger helper with me. He was green in some areas, but hungry to learn. I could have done what a lot of leads do: bark orders, let him haul trash, and keep all the "real" work in my own hands so it would go "faster."

Instead, I slowed down just enough to start teaching.

I showed him how I looked at a room before I ever brought a roll in. I explained why we were snapping a line where we did. I walked him through what I was seeing when I checked a seam for pattern and backing, not just surface. I let him make a few mistakes under my supervision, then made him fix them the right way.

We didn't gain any time that day. If anything, we spent a little extra. But when we left, the work was solid and something else had changed: I wasn't the only person in the world who knew how that job went together.

Driving home, I realized I didn't actually want out of flooring. I was just tired of being the only engine pulling the whole train. I still liked the craft. I still liked making something right under people's feet. What I wanted was a different role in it as the years went on—more teaching, more leading, more using my head instead of just my joints.

Checklist – Staying in the Game for the Long Haul

USE THIS AS A QUICK, honest check occasionally.

Body

- Am I using decent knee pads most days, not just when I remember?
- Am I using carts, dollies, and help for heavy loads instead of ego-lifting?
- Am I protecting my lungs and ears on grind-and-glue days?
- Do I stretch or move enough that my back isn't seizing up daily?

Mind

- Am I saying "yes" to jobs that crush me more than they pay me?
- Do I have any real time when I'm *off* and not answering work stuff?
- Do I still feel some pride and interest in the craft—or just resentment?

Crew

- Do I have at least one person I'm truly training, not just using?
- Have I clearly explained what "our standard" is, instead of assuming they know?
- Do I deal with bad fits instead of carrying them forever?

Systems & Future

- Do I have a few written checklists my crew uses?
- Am I building relationships with the kind of clients I want more of?
- Do I have any plans for doing less crawling and more directing over time?

You don't have to fix all of this in one week.

Just pick **one** item in each section and move it one step forward:

- Better knee pads
- One clearer boundary
- One honest conversation with a helper
- One small system you write down

Do that repeatedly, and one day you'll look up and realize:

You didn't just survive this trade.

You built a career—and a life—that was worth the miles on your knees.

Chapter 14 – Building a Name That Sticks

Your tools matter. Your stretches matter. Your seams and prep matter.

But in the long run, what really travels is your **name**.

Most people who walk your floors will never know it. They'll just know whether the carpet feels solid, quiet, and safe—or loose, noisy, and tired. The folks who *do* remember your name are the ones who live with your work every day:

- The front desk who stops getting trip complaints.
- The facility manager who stops chasing bubbles and ripples.
- The GC who stops holding their breath every time "carpet" comes up on a punch list.

They may not talk about "tension," "seam tape," or "subfloor tolerances," but they will say things like:

"If they're on the job, the floor will be right."

That sentence is your brand. This chapter is about building that on purpose.

Professional Habits That Write Your Story

YOUR REPUTATION DOESN'T start with logos or websites. It starts with boring, repeatable habits that other people quietly notice.

Show up when you say you will.

Nothing wrecks trust faster than ghosting a job or drifting in late with excuses. Call ahead if something truly changes. Being reliable is more impressive than being a genius.

Finish what you start.

Punch lists, callbacks, tiny seams that bug *you* more than the client—how you handle those is what people remember. A quick, clean follow-up tells them, "If something ever goes wrong, they won't vanish."

Leave it cleaner than you found it.

Scrap picked up, blades contained, pad pieces gone, vacuum tracks visible. A room that feels "ready to use" when you leave tells clients you care about more than just your square yards.

Be the calm one.

Jobs get loud. Schedules shift. Other trades talk sideways. The installer who can stay steady, explain what's happening, and solve problems without drama becomes the one people fight to keep around.

Make Yourself Easy to Find

YOU DON'T NEED A MASSIVE brand. You do need to be findable.

Keep it simple:

- A clear company name that's easy to say and spell.
- A basic logo on shirts, hats, or your truck—even if it's just your name cleanly laid out.
- One current place online (website, Google listing, or social page) with photos and contact info.

When someone says, "Who did your floors?" you want your client to be able to say, "These guys—here's their site," not, "I think his name was Mike... from somewhere?"

Your visual identity isn't about looking fancy. It's about letting your good work have somewhere to point.

Word-of-Mouth and Staying in Their World

THE BEST MARKETING in this trade is still one person telling another, “Call this installer.” You can't force that—but you can make it easier.

After a solid job, don't just disappear.

- Send a short text a few weeks later:

“Just checking in—floor still treating you right?”

- When you see they've added a wing, opened a new location, or changed managers, send a quick congratulations and reminder that you're there if they need you.
- For your favorite repeat clients, a simple holiday card with a real signature goes a long way.

You're not pestering them. You're reminding them you exist *before* a flooring issue lands on their desk.

If they're happy with your work, it's also fine to say:

“If you know anyone else who needs floors done right, feel free to pass my name along.”

Simple, honest, and respectful. That's all you need.

Price Like Your Name Matters

WE COVERED PRICING nuts-and-bolts earlier in the book. Now, think about price as part of your **story**.

If you're always the cheapest, people assume you're:

- Desperate,
- Cutting something they can't see, or
- Not fully aware of your own costs.

If you're in the upper-middle of the pack and can explain clearly what's included—prep, protection, stretch, follow-up—your price starts to sound like *insurance*, not a gamble.

You don't have to be the highest. You do have to charge in a way that matches the standard you actually want to work at. Never price a job so low that you're tempted to cut corners just to survive it. That's how reputations rot from the inside.

If someone only cares about the bottom number and nothing about the quality, fit, or longevity, let them shop somewhere else. Your name is attached to that floor long after the invoice is buried.

Think in Years, Not Weeks

SHORT-TERM THINKING says, "Can I get through this job and get paid?"

Long-term thinking asks:

- "If they call me back in three years, will I be proud or embarrassed?"
- "If another installer pulls this up, what will they assume about me?"
- "If this client wants another job at this price and this standard, do I want more of that or less?"

A name that sticks isn't built on perfect jobs. It's built on:

- Honest pricing,
- Clean follow-through,
- Owning mistakes, and
- Walking away from work that would force you to be someone you're not.

Every project is a vote for the kind of installer you're becoming.

Job Note – The Floor I Still Think About Years Later

EVERY INSTALLER HAS a few jobs that live in their head long after the check clears.

One of mine wasn't huge or famous. No casino, no luxury hotel, no magazine photos. Just a mid-size lobby and corridor in a building most people would drive past without noticing.

What made it stick wasn't the square footage. It was how we treated it.

We pushed for the right prep instead of "good enough."

We fought for a layout that respected the pattern and the traffic, even though it meant more cutting.

We gave seams time to cool and stretches time to settle instead of rushing to move on.

We cleaned that place like we were moving in ourselves.

Years later, I ended up back in that building for a meeting. Same carpet. Still flat. Still tight. Seams still calm under the lights.

I watched people walk across it.

Nobody tripped.

Nobody stared at the floor.

Nobody knew or cared who installed it.

But I knew.

Driving away, I realized: this is the kind of floor I want my name tied to—quietly doing its job for years, in a building nobody writes articles about, under people who never learn who I am.

That job will never be on a billboard. But it's part of my brand whether anyone sees it or not.

Checklist – Ten Ways to Make Your Name Mean Something

USE THIS AS A QUICK audit every few months. Be honest; nobody else has to see it.

Habits

- Do I show up when I say I will—or clearly communicate when I can't?
- Do I finish punch lists promptly instead of letting them drag on?
- Do I leave jobs cleaner than I found them?

Communication

- Do my quotes and emails sound like a professional, not a rushed text thread?
- Do I set expectations about noise, dust, access, and cleanup *before* the job starts?
- Do I follow up at least once after completion?

Visibility

- Do I have recent photos of my work that I'd be proud to show anyone?
- Is there a simple, current place online where people can find me and see what I do?

Boundaries & Pricing

- Do my prices honestly reflect the standard I want to work at?
- Am I willing to walk away from jobs that would force me to cut corners or undercut my own name?

If most of those answers are "yes," your name is already carrying weight when you're not in the room.

Pick one or two to fix now. Names are built over years, not weeks.

You don't need to be famous.

You just need the right people—owners, GCs, staff, other trades—to think one clear thought when your name comes up:

"If they're on the job, the floor will be right."

That's a name that sticks.

About the Author

T.J. Spracklin has spent more than three decades on his knees with a knife in his hand, installing floors in hotels, casinos, offices, showrooms, and homes across America. From high-pressure overnight corridor flips to delicate pattern work under brutal sunlight, he's made most of the mistakes this book is trying to help you avoid.

As a master installer and storyteller, T.J. writes for the people who actually crawl the rooms and carry the rolls—the helpers, leads, and small-shop owners whose names rarely go on the plans but whose work holds the building together. When he's not laying down carpet or coaching other installers, he's building blue-collar sagas and playbooks that treat the trades with the respect they deserve.

Also by T.J. Spracklin

Fiction – Seams of Power Saga

- *Seams of Power – Book 1: Under Every Step*
- *Seams of Power – Book 2: Hidden Network*
- *Seams of Power – Book 3: Anchored Ground*
- *The Last Seam: Carpet Underworld*

Trade & Craft

- Seams of Power – Book 4: *Installers Playbook*
- *Laying It Down Right: The Master's Guide to Carpet Installation* (this book)

For more books and updates, visit spracklinbooks.com

www.ingramcontent.com/pod-product-compliance
Lightning Source LLC
LaVergne TN
LVHW090320210926
848480LV00001B/310